Stealing into Print

Stealing into Print

Fraud, Plagiarism, and Misconduct in Scientific Publishing

MARCEL C. LAFOLLETTE

University of California Press

BERKELEY   LOS ANGELES   LONDON

University of California Press
Berkeley and Los Angeles, California

University of California Press, Ltd.
London, England

Copyright © 1992 by The Regents of the University of California

Library of Congress Cataloging-in-Publication Data

LaFollette, Marcel Chotkowski.
   Stealing into print : fraud, plagiarism, and misconduct in
scientific publishing / by Marcel C. LaFollette.
     p.   cm.
   Includes bibliographical references and index.
   ISBN 0-520-20513-8
   1. Science publishing—Moral and ethical aspects. 2. Publishers
and publishing—Professional ethics. 3. Technical writing—Moral
and ethical aspects. 4. Authorship—Moral and ethical aspects.
5. Scientists—Professional ethics. 6. Plagiarism. 7. Fraud.
I. Title.
Z286.S4L33   1992
179'.9097—dc20   91-41669
                                          CIP

First Paperback Printing 1996

Printed in the United States of America

1 2 3 4 5 6 7 8 9

Contents

Acknowledgments

Many people have made this book possible. Some of them are named below; others, for reasons of confidentiality or privacy, are not. All offered kindness in sharing details of their personal experiences or bringing new material to my attention, intellectual generosity in listening to or critiquing my ideas, and unfailing enthusiasm for a project that seemed to go on forever.

For suggestions, inspirations, and support at various stages I thank: Joseph Brenner, Nancy Carson, Larry Cohen, Ellen Chu, Catherine V. Chvany, Patricia Garfinkel, Harold P. Green, Melvin Kranzberg, Wil Lepkowski, Nathan Reingold, and William A. Thomas.

Extensive critiques by Lawrence Badash, Daryl E. Chubin, Elizabeth Knoll, and Drummond Rennie were invaluable in the last revisions. Daryl Chubin, in particular, has been a steadfast, generous colleague and good friend during stormy weather.

The Center for International Science and Technology Policy at The George Washington University, and its director, John M. Logsdon, provided a congenial home at a crucial stage, and the other inhabitants of the Center, Kimberly Lutz, Paul McDonnell, Robert W. Rycroft, and Nicholas Vonortas, have made it a pleasure to be there. Thank you all for encouragement and good cheer.

No list of thanks would be complete without mention of the tuneful triumvirate—Bob, Elvis, and Sid.

And, finally, my husband, Jeffrey K. Stine, offered unfailing, unconditional love, patience, and good humor. Come give me your hand, dear—let's go for a long walk and talk about *your* work for a change.

The National Science Foundation, under NSF grant No. RII-8409904, funded developmental research that led to this book; Cynthia Closkey served as research assistant on the NSF grant. No other federal or private grant supported the remainder of the writing; the opinions and conclusions in the book represent those of the author, not necessarily of any institution with which she is or has been affiliated.

1   When Interests Collide
Social and Political Reactions

For it is a matter which concerns every author personally . . . that his own proper labors, the hard-earned acquisitions of his own industry and intellect, should not be pilfered from him, and be represented by another as his own creation and property.

—*The Lancet*[1]

It's sad if people consider them high standards. They should be ordinary standards.

—Margot O'Toole[2]

Truth swings along a wide arc in the late twentieth century. In politics and popular entertainment, people accept deception as commonplace— campaign image making, video synthesis, and the trompe-l'oeil of fantasy parks seem normal parts of life. The trust that society places in science, however, traditionally assumes different standards, including assurances of authenticity and accuracy in all that science does or recommends.

The strength of this faith in science's intrinsic truthfulness helps to explain the roots of the current political controversy in the United States over "scientific fraud." Forgery, fakery, and plagiarism contradict every natural expectation for how scientists act; they challenge every positive image of science that society holds.

The sequence of events that led to federal regulations (and the heated political debate surrounding them) initially attracted little attention, from journalists or even from the scientific community. In the 1960s and 1970s, the most outrageous examples of faked data or plagiarism were dismissed as aberrations, as unrepresentative of the integrity of scientists overall.[3] Then, in the 1980s, cases began to surface at Harvard, at Yale, at state universities, and in several different fields. Scientists employed at the best places, who had attracted hundreds of thousands of dollars in federal research funding, and who had published in top-ranked journals were accused of forging, faking, or plagiarizing their way to success. Reputable

1

laboratories and schools were accused of not just indifference but coverup. Soon, the congressional committees charged with overseeing U. S. research and development (R&D) programs began to investigate why the management and evaluation systems created to monitor research and research communication had apparently failed to detect or prevent misconduct. These monitoring systems had served science well for decades; but the basic concept of peer review—the reliance on experts to review journal manuscripts before publication—rested on tacit assumptions about the honesty and truthfulness of all participants in the process, and these were exactly the old-fashioned virtues that deception mocked.

Now the klieg lights are on. What began as private discussions among professionals stands unshielded in the spotlight of national politics. Moreover, controversy has spread beyond the relatively few institutions that have been directly implicated and has spilled from the hearing rooms of Capitol Hill to the pages of morning newspapers.

No other science policy issue of our time contains such potential for altering the who, what, how, and whether of the production and dissemination of scientific knowledge and for affecting the rights of individual authors and publishers. This book focuses on one segment of that enormous issue—how the controversy affects communication practices and policies in the journals that disseminate the results of scientific research—using it as a window on the larger debate. What does the discovery of unethical conduct say about current systems for evaluating and disseminating research-based knowledge? And how may those systems be changed by efforts to detect, investigate, mitigate, and prevent wrongdoing?

The issue has a number of unusual characteristics worth noting at the outset. One is the level of emotion present, explicitly and implicitly, in almost every discussion. One can point to similar psychological reactions to forgery or plagiarism in other fields of scholarship and creativity. In reacting to literary plagiarism, for example, Edgar Allen Poe once wrote that he was "horror-stricken to find existing in the same bosom the soul-uplifting thirst for fame and the debasing propensity to pilfer. It is the anomaly, the discord, which so gravely offends."[4] Identical disbelief greets the news that scientists have seemingly faked data for the sake of a few more publications. Deception seems anomalous. Time and time again, shocked co-workers have been heard to mutter such phrases as "all of us thought very highly of this young man. . . ," and then they shake their heads and turn away in disgust.[5]

Another significant aspect of the issue is that the discovery of unethical or illegal conduct contradicts scientists' self-images, their beliefs about how "real" scientists act. As a consequence, the community often engages in wholesale denial of the problem's significance. There is deep-seated disagreement over elementary procedural points, leaving investigators uncertain about whose standards to apply in assessing guilt or innocence. Because congressional interest in research fraud coincides with other prominent and highly controversial science policy debates, such as those over resource allocation, priority setting, and university overhead, and with the perception that standards are lax in all professions, scientists who are accustomed to deafening applause suddenly feel beleaguered by unwanted criticism.

Scientists also tend to perceive themselves as defenders of objectivity in an irrational world, struggling to rebuff superstition and pseudoscience. Popular culture reinforces those self-images when it assigns scientists the role of seekers, determiners, and guardians of truth.[6] Deception and fraud shatter the glistening surface of these images.

The discovery of fraud among a nation's intelligentsia also provokes reconsideration of its core values, revealing where those values may be self-contradictory. Entrepreneurship, independence, and ambition, for example, are important traits in American culture. As a result, the nation's scientists have been encouraged to stretch their imaginations and energies to advance the frontiers of knowledge, encouraged to engage in intense competition as a matter of course. Ironically, Americans also advocate fair play, and they expect winners to be humble and generous, to share credit where credit is due, and not to steal (or award) credit falsely. The greedy overtones of fraud and plagiarism seem intrinsically out of character for scientists, seem unacceptable behavior for society's heroes.

In addition to these emotional and cultural components, the political controversy has been distinguished by unrelenting disagreement over basic definitions. Is "scientific misconduct" an ordinary social crime committed by deviant individuals, does it represent an indicator of slipshod research management, or is it a symptom of widespread social immorality? Whose standards should be applied and what specifically should be permitted or prohibited? The failure to develop widely accepted definitions of expected research conduct has resulted in years of unproductive debate and unworkable *ad hoc* policies.

The political dispute over research fraud taps into larger issues of national policy. The political framework adopted for post-1945 organization of U.S. science assumed that scientists were trustworthy and that their technical expertise was always reliable. American researchers have enjoyed considerable autonomy in directing their own research agendas and procedures compared to scientists elsewhere in the world. Each report of suspected research fraud tests the soundness of this approach; each proof of wrongdoing further erodes political support for unilateral autonomy and raises more calls for accountability. The combined attention of congressional committees, federal science agencies, and the press has thus slowly shoved the focus from initial concern about personal conduct (and a preoccupation with remedial moral counseling) to foundational questions of who sets the policies, public and private, for scientific research and communication. And that debate will have implications for the international systems of scientific communication, including commercial and nonprofit periodical and book publishers, secondary indexing and abstracting services, libraries, and similar information processing organizations around the world.

The recent suggestions that journals afford ideal locations for detecting, investigating, and rooting out fraud must therefore be carefully considered. One of the glories of U.S. science (and perhaps one of the keys to its success) is the extent of freedom of communication within the research system. In theory, any person is free to publish and speak about science, no matter how innovative, radical, or ridiculous the idea because, although peer review is widely regarded as a means of *authenticating* published science, it was never intended as a method of scientific censorship. Mistakes or errors within the system may simply represent a price paid for a free market in scientific information and, if the risk is unacceptable, then the alternatives should be carefully debated. Locating an appropriate balance between the need to prevent and punish the unethical practices described in this book and the need to preserve individual freedom of expression will require cooperation among the scientific community, the publishing community, and government policy makers. This book recounts how their various interests continually intertwine—and sometimes collide head on.

SETTING THE SCENE

Even under the best of circumstances, allegations of illegal or unethical professional conduct stimulate a maelstrom of conflicting reactions. Colleagues and friends alike react emotionally—sometimes with denial, sometimes outrage. Everyone has opinions. No one stays quiet.

To illustrate the types of issues and the miasma of conflicting interests and attitudes that typically surround misconduct allegations in the context of scientific publishing, I begin with descriptions of two episodes from the nineteenth century and two from more recent times. Research procedures and protocols may have changed over the course of the years, but human failings remain constant.

To Read, to Translate, to Steal

It is routine practice in research and scholarship to award special prizes for particular writings and then, as part of the reward, to publish the essay. Such a process took place in 1852, when the British Royal College of Surgeons awarded its Jacksonian Prize to an essay by Henry Thompson and published his essay in London in 1854. Two years later, in Paris, scientist José Pro presented an "original" essay on the same subject to the Societé de Chirurgie, which awarded him their highest honor. Twenty-three of the twenty-six quarto pages of Pro's essay was later proved to be a "literal translation" of Thompson's prize essay.

British scientific reaction was characteristically understated in its outrage. *Lancet*, a medical journal published in London, printed extracts from the two works side by side as indisputable evidence of the plagiarism; the editors prefaced the columns with sarcastic praise for "M. Pro's undoubted merits as a translator."[7] In biting criticism of the French medical establishment, *Lancet* wrote: "M. Pro, in perpetrating one of the most shameless and extensive plagiarisms which has been brought to light for many years, has paid a great compliment to their real author, and . . . the Societé de Chirurgie, by conferring their highest distinction upon the supposed original observer, have quite unwittingly confirmed it."[8]

Lancet recognized that the case had implications far beyond simple plagiarism, especially in the world of research where scientists must trust other scientists to give them due credit. The editorial pointed out that the wrong was not just Thompson's concern alone; it was shared by the entire community. Every author, the editors wrote, desires "that his own proper labors, the hard-earned acquisitions of his own industry and intellect, should not be pilfered from him, and particularly that those acquisitions should not, in some more or less distant parts of the world, be represented by another as his own creation and property."[9]

Like many similar episodes, this one was later used as a sort of Aesop's fable of scientific morality, to introduce an account of similar plagiarism in the United States.[10] During the winter of 1865–1866, George C. Blackman, a physician in Cincinnati, Ohio, received a copy of a prize-winning medical essay that had just been published in France. In 1862, the Academy of Medicine of Paris had proposed a history of research on *Ataxia Locomotor Progressive* as the subject of a prize essay, and Paul Topinard, a member of the Societé Medicale d'Observation, had entered and won the competition by adding original clinical observations on the disease to his analysis of data on 252 previously reported cases. Like countless scientists before and since, Blackman loaned his copy of the Topinard essay to a colleague, Roberts Bartholow, who had "professed a desire to study the subject."[11]

A few months later, an article by Bartholow, titled "The Progressive Locomotor Ataxia (*Ataxia Locomotor Progressive*)," appeared in an issue of the *Journal of Medicine*, published in Cincinnati. In a note printed at the bottom of the article, Bartholow stated that he had made "liberal use" of Topinard's essay. In fact, his debt to Topinard was more than token. After giving the details of an Ohio case that he had studied, Bartholow devoted the remainder of the article to what was actually a *literal* translation of the French essay.

When the publisher of the Cincinnati journal discovered the apparent plagiarism, he asked Bartholow to address the accusation in part two of the series, which was scheduled to appear in the next issue. Bartholow complied, stating:

> It having come to my knowledge that some persons consider my reference insufficient to the work of Dr. Paul Topinard (*De L'Ataxie Locomotor Progressive* etc.) I beg to inform the readers of the Journal that my intention was, by the term, "liberal use of the Essay" to express the idea of a synopsis of such parts as suited my purpose in the preparation of my article. As the work is a voluminous one, it would be impracticable to present a translation in the number of pages allotted to me. The present, as the former article, is a synopsis chiefly of the views of Topinard.
>
> ———R. B.

George Blackman was apparently unsatisfied with Bartholow's explanation and decided to publicize further what he regarded as his colleague's wrongdoing.[12] Like the *Lancet* editors a decade earlier, Blackman became a "nemesis," attempting singlehandedly to punish unethical behavior through dogged publicity. In 1868, he privately published a pamphlet in which he outlined the allegations, drawing on a description of the Thompson/Pro story for comparison. "As preliminary to the main object of this paper," Blackman wrote, "we will briefly refer to a literary transcription of 'extraordinary peculiarity,' which occurred in the great metropolis of France, and then proceed to show how the said transcription has even been eclipsed by a writer of the Queen City."[13]

Blackman facetiously praised his colleague's abilities as a translator and then accused Bartholow of having attempted to dispel suspicion by acknowledging a "liberal" use of "a celebrated French Prize Essay on the same subject," when in truth the use was *literal*. Blackman juxtaposed columns from the two Bartholow articles and the original prize essay, demonstrating that not only the texts, but also their openings and sub-heads, were identical in many places. Although Blackman acknowledged that the third part of Bartholow's article (published in June 1866) bore "less evidence of being 'a debasing *translation* and unscrupulous assumption' of knowledge . . . ," he remained convinced that Bartholow's original intention was to deceive. Blackman's condemnation of the character of the accused was scathing; his desire to expose him, intense; and his method, publicity.

To Republish

Such attitudes toward theft of ideas and such actions to publicize it are not simply quaint artifacts of the nineteenth century. They typify attitudes within science that are expressed vigorously today, sometimes behind the scenes and sometimes in "corrective action" taken to publicize unethical behavior. A typical episode occurred in the 1980s, when news reports began to describe a "rash of what appears to be piracy . . . in the scientific literature" involving a Jordanian biologist, E. A. K. Alsabti, and several journal articles he had published.[14]

Alsabti apparently took articles published in one country (usually in small-circulation journals) and published them under his own name in journals published elsewhere. He also allegedly retyped previously published articles and then submitted them to other journals. In one instance, he reportedly obtained a preliminary version of a manuscript which he successfully published under his name before the original author could get the article into print.[15]

Alsabti did not have a permanent faculty appointment in the United States and, during the period in which he engaged in these activities, he moved frequently. Because none of the universities or laboratories in which he had worked was willing to conduct a formal investigation, several scientific journals, most notably *Nature* and the *British Medical Journal*, took up the banner. They independently addressed the problem by publicizing the wrongdoing—printing paragraphs from the original texts side by side with Alsabti's articles, along with editorial comment on his "piracy."[16]

The journals' actions represent the only direct confrontation of Alsabti's behavior by any institution at the time. Alsabti neither admitted to plagiarism nor retracted the articles in question, although (after a legal battle that stretched for years), the Commonwealth of Massachusetts was able to revoke his license to practice medicine, citing the alleged plagiarisms.

Nature and the *British Medical Journal* aggressively expressed community outrage in one of the few ways possible, given the lack of available mechanisms or procedures for bringing charges on plagiarism and also given Alsabti's steadfast assertion of innocence. The journals pilloried Alsabti just as the *Lancet* and George Blackman had done in similar cases almost a century earlier. By publishing comparative texts, the editors led readers to act as a "jury of peers" and evaluate the actions for themselves. They also forced action when the normal institutions for doing so either could not or would not respond.

To Publicize, to Analyze

Scholarly and scientific journals, then, may unwittingly aid in disseminating plagiarisms or descriptions of faked data or artifacts; they also attempt to correct these lapses by giving publicity to wrongdoers and by serving as the forums in which scientists can discuss or debate issues of professional ethics. One of the most troubling examples of the latter functions took place in the 1980s. The resulting controversy drew mass media attention, provoked congressional hearings, and emphasized dramatically that editors and publishers could not stand back and wait for universities or government agencies to sort out the disagreements on their own. The publishing community not only had a stake in the debate but, one way or another, it was being pulled into the political controversy, whether it wanted to be there or not.

This episode began in the laboratories of one of the most prestigious medical schools in the world. John Darsee was a promising young researcher at the Harvard Medical School, working under the mentorship of a top cardiologist, Eugene Braunwald. During his previous graduate work at Emory University, and later at Harvard, Darsee had appeared to his professors, mentors, supervisors, and most co-workers as a typical, bright, energetic, and ambitious post-doc. His research, part of an important multi-institutional project, was funded by the National Institutes of Health (NIH); when the larger project concluded, several years after the Darsee controversy died down, its final results were, in fact, widely praised. Scandal erupted in 1981 when Darsee was accused of manipulating, inventing, or otherwise compromising the integrity of data reported in over a dozen coauthored papers and over fifty abstracts based on his Harvard cardiology research.[17]

Darsee's dismissal and the sensational details of the allegations attracted considerable press coverage. The event seemed to imply that fraud could occur even at the "best" universities, where integrity and truth were emphasized in school mottos and inscribed on the very buildings in which the scientists worked. As the official conclusions of Harvard University, Emory University, and NIH investigating committees were released in

1982 and 1983, one issue remained unanswered, however: exactly what role had been played by Darsee's coauthors? Darsee had published the faked data in many articles that also bore the names of senior faculty and junior postdoctoral researchers, both at Emory and at Harvard, yet these scientists had apparently failed to detect the fabrication and had willingly cooperated as coauthors.

In retrospect, it may seem remarkable that no one questioned Darsee's phenomenal rate of production or asked to see his data, but such a level of trust among co-workers is normal and, indeed, encouraged among scientists. As it became known that many of Darsee's coauthors were, in fact, "honorary"—that is, they had not actually participated in the research or in the writing of the questionable articles, yet had allowed their names to be listed on the papers—embarrassing criticism of them began to appear in the press. In editorials and in letters to scientific journals, outraged scientists questioned the wisdom of prevailing standards for authorship, the quality of editorial review at major journals, and the growing practice of "nonparticipating" coauthors. The Darsee episode began to spark intense debate about who should receive credit for collaborative research, and why.

The practice of honorary coauthorship particularly disturbed two biologists, Walter W. Stewart and Ned Feder, who worked in the National Institute of Arthritis, Diabetes, and Digestive and Kidney Disease (NIADDKD), part of the NIH. To Stewart and Feder, any coauthor who accepted credit for an article should accept responsibility for its accuracy (and therefore accept blame if the article was later found to be flawed). As a sideline to their own research, the two biologists began to analyze "the Darsee coauthors" and to write an exposé of the practice of honorary coauthorship in general.

The early scientific work of Stewart and Feder is highly regarded, but they are biologists, not social scientists; they lack training in the theories and methodologies of sociology, and the early drafts of their manuscript, couched in the language of the social sciences, reflected their unfamiliarity with its pivotal literature and interpretations.[18] Had they phrased their remarks as general editorial comments by two biologists on the issue of coauthorship, without purporting to be presenting data collected as part of social science research, their manuscript might well have received a normal review and appeared, with little fanfare, in any number of journals. Instead (and perhaps because it was the world they knew best), they attempted to cast their article as objective "science." They focused on a specific group of John Darsee's coauthors, identifying them directly by name or indirectly by citation to the articles in question. They used blunt language in condemning the coauthors' actions. And they doggedly attempted to publish their article in the most eminent scientific journals in the world.

Their efforts to publish quickly became a *cause célèbre*.[19] Stewart and Feder first gave a draft to John Maddox, editor of *Nature,* in 1983; by early in the next year, the manuscript had been revised repeatedly. Some time after Maddox gave a version to one of the "Darsee coauthors" named in the manuscript, a lawyer for Eugene Braunwald (the most prominent of the scientists named, and the head of the lab in which Darsee had worked) wrote to Stewart and Feder, claiming that the draft contained "defamatory statements." The attorney had similar discussions with representatives of *Nature.*

Over the next four years, while they continued to seek publication, Stewart and Feder were threatened with lawsuits claiming that their manuscript contained libelous statements.[20] The coauthors who threatened to sue did not condone or excuse Darsee's behavior, but they were convinced that publication of the article would damage their professional reputations. One attorney began to warn journal editors that even circulating the manuscript to peer reviewers, much less publishing it, would draw legal complaints.[21] Such threats raised anew the prospect that private disputes between scientists might be propelled into the courts and they pushed the issues to a new stage.[22] Heretofore, most science editors, authors, and referees had—with good reason—tended to regard editorial decision making and the peer review system as closed, protected activities, more or less like professional conversations among colleagues.

The accusation of libel further delayed publication; by December 1984, after more revisions, Stewart and Feder had received NIH approval on a "final" version, but Braunwald's attorney continued to claim that the text contained libel.[23] Perhaps out of pique, perhaps just frustrated at the delays, Stewart and Feder began to circulate throughout the scientific and science policy communities a packet containing the draft manuscript and copies of what was already a voluminous correspondence on the matter. Maddox requested another round of revisions, and the biologists withdrew the article. The second phase began.

Stewart and Feder passionately clung to the hope that their manuscript would be published in a science journal because they believed that discussions of this ethical issue should be confined to scientists. So they next formally submitted the manuscript to *Cell,* beginning an unsuccessful review process that lasted about eight months. Hedging their bets, they also began to contact editors in the science studies community.

The biologists' tireless efforts and the reactions they stimulated seem even more extraordinary in retrospect. Journalist Daniel S. Greenberg, a long-term observer of science and politics, later called the manuscript "a scientific Flying Dutchman, unable to find a place of publication, despite praise from reviewers and expressions of regret from editors."[24]

Stewart and Feder first called me in 1985, in my role as editor of *Science, Technology, & Human Values (STHV)*. That experience, although mild by comparison to that of those directly caught in the legal controversy, nevertheless provides a firsthand example of how even ordinary journal publishing can contain multiple conflicts of interest.[25] Nothing is straightforward. Where no formal rules govern publishing policies or procedures, ambiguity reigns, for better or worse.

As a formal policy, *STHV* discouraged "double submission" of articles, that is, the practice of submitting an article simultaneously to two or more journals without the editor's knowledge or permission; potential authors were asked to attest, upon submission of an article, that it was not under review or had not been submitted elsewhere. Stewart was convinced, however, that *Cell* would soon reject the manuscript and he implied that, if *Cell* did reject the wandering manuscript, it might next land at *STHV*.

Like most academic journals, *STHV* relied on delicate political coalitions for its survival. In those years, the journal was cosponsored by academic programs at Harvard University and the Massachusetts Institute of Technology (MIT). It was co-owned by the two universities and published by the commercial firm John Wiley & Sons. This coalition created a delicately balanced situation. The universities and the publisher supported the editor's freedom to make all decisions on content; in return, they expected the editor to make prudent choices that would not violate the law, would not jeopardize the reputation of the journal or its sponsors, and would not impose undue financial burdens on the sponsors or the publisher. The *STHV* editorial advisory board, like those at most small journals, was consulted about matters of long-term policy, but not about the daily editorial operations or particular manuscripts (unless a board member was an expert on the topic). This was clearly no ordinary manuscript, however. Even to circulate it to peer reviewers meant attracting controversy and, possibly, legal action. "Wise counsel" was in order. So, I consulted the publisher, the university attorneys, and a few trusted members of the board. Their reactions and recommendations represented typical opinions about these matters—appropriate, honorable, divided.

The publisher, for example, reacted with strong support for moving forward with review and publication and informally encouraged me to do so. Any controversy over this type of ethical issues would have attracted favorable publicity and probably increased circulation, of course; but John Wiley & Sons is also one of the oldest, most respected scientific and engineering publishers in the United States. Its executives apparently bristled at the idea that the threat of a libel suit should ever be allowed to influence expert peer review or to hobble editorial freedom.[26]

According to the contract between the universities and the publisher, the academic sponsors warranted that all content transmitted to the publisher for typesetting was (among other things) free of libel. The universities, not Wiley, therefore, would have had to bear the legal and financial responsibility for responding to a lawsuit. Yet, in this case, the Office of the General Counsel at Harvard, which normally provided legal advice to the journal, was also advising at least two of the Harvard faculty members involved in the Darsee case. A clear conflict of interest prevented involvement of the counsel's office. Had *STHV* actually become embroiled in a legal controversy, lawyers for MIT and Harvard would have had to sort out these obligations and agree on how to protect their joint as well as their individual interests. Fortunately, that complicated process was never tested.

Conversations with the journal's advisory board members brought mixed recommendations. Some urged me to proceed with peer review. Others urged just as strongly that the matter be dropped for fear that "bad publicity" might endanger *STHV*'s survival or my professional career. Even rejecting the manuscript on the basis of inadequate scholarship would have attracted criticism. Almost everyone suggested that I retain an attorney, an expensive prospect for a junior faculty member who was then serving as the journal's editor without compensation.

Thankfully, no journal had to take the legal risks in order to publish the article. On 26 February 1986, Stewart and Feder testified at a hearing before the House Committee on the Judiciary. Its Subcommittee on Civil and Constitutional Rights was then investigating "the 'chilling effect' of libel threats on editors' willingness to publish controversial material."[27] In April, the *New York Times* reported extensively on Stewart and Feder and their findings.[28] And on 14 May 1986, the House Committee on Science and Technology held a hearing on the controversy surrounding the manuscript and the article was published in the *Congressional Record*, thereby effectively eliminating the threat of libel.[29] Here, the link between the professional conduct of scientists, the discussion of ethical issues in journals, and the broader issues of national science policy were joined. The House Committee's staff recognized that publication practices were linked to a changing climate for R&D management, to an atmosphere of ever more frantic competition among young researchers, and to the intellectual integrity of the knowledge base.

Nature published the paper in 1987, accompanied by commentary from

Eugene Braunwald, seemingly ending the saga.[30] One headline declared "Stewart and Feder (Finally) in Print," but, as I shall discuss in chapter 6, this episode was not the last controversy to engage the interest of Walter Stewart and Ned Feder.

Courage is easy to summon in retrospect. I cannot truthfully say what I would have done if I had been forced to decide whether to proceed with formal peer review and to risk damaging a journal I had worked so hard to build. The episode left a sour taste in many mouths. And it threw an embarrassing spotlight on the weaknesses in editorial policies, revealed how difficult the choices could be, and showed how inadequately prepared the journals were to meet them. We were, at the least, unaware of the common ground. The same laws that protect or punish mass-market newspapers and magazines apply to scholarly journals. And the same shifting political environment that influences research funding and practice also affects its communication.

MULTIPLE EXPLANATIONS

Anecdotes can show how roles play out in particular episodes, but understanding the broader issues at stake requires a look at the historical, social, and political context in which the episodes take place. For the political controversy over scientific misconduct, a significant aspect of that context is the persistence of certain explanations for what is happening. Foremost among these is the unwillingness to accept that any scientist, any *real* scientist, could deliberately fabricate or misrepresent data and then describe that data in scientific articles as if it were real. Framed as questions repeatedly posed by many people in many forums, these rationalizations form an intellectual grid on which to impose the history of the political controversy.

How Much Is There?

Like a dog nipping at a runner's heels, this question has nagged the analysis and discussion of research fraud from at least the 1970s. Is there increased incidence of wrongdoing, are the standards for behavior getting higher, or is there just better detection? Is the potential for unethical conduct inherent in science for all time or inherent in how research is conducted today?

Deceitful appropriation of credit and ideas is not a new problem, of course, either for scientists or any other creative professionals. A cursory glance at history tells us that. Social norms and expectations (and, hence, the definitions of what constitute violations of standards) may change over time within individual professions or research fields, but outright fakery and unattributed theft of another's ideas have never been considered appropriate behavior.

The degree of formal condemnation, sanction, and punishment of particular actions *has* varied, however, over time and among professions. Once "authorship" began to carry the potential for profit, then plagiarism became a matter of concern to all authors and publishers; once "antiquities" began to be perceived as objects of value for commerce as well as scholarship, then forgery and purveying of fakes by dealers, archaeologists, and private collectors became activities worth the risk of detection.[31]

Through the years, science, like every other human occupation and profession, has attracted its share of crooks and criminals, as well as its heroes and individuals of integrity.[32] Nevertheless, even though the theft of scientific ideas may appear on the increase, no systematic assessment of the ratio of honest to dishonest people or activities in science has ever been made or indeed could be made because practices of attribution among disciplines, over time, even among contemporaneous institutions, are simply not comparable. In the Renaissance, for example, "authors and sources were cited only in case of refutation."[33] Even Leonardo da Vinci was accused of appropriating ideas and adopting engineering designs from his contemporaries. One likely target of such theft, Francesco di Giorgio Martini, a fifteenth-century Sienese artist and architect, published his inventions only reluctantly because he was convinced they would be stolen by "ignoramuses," as he called them, who sought to "adorn themselves with the labours of others and usurp the glory of an invention that is not theirs."[34]

For many critics of modern science, the fact that deception or misappropriation of ideas may be an age-old problem provides insufficient reason to dismiss current episodes as mere dots in a continuum. To them, today's ethical failings among scientists seem qualitatively different. Instead of blaming science, however, they blame its social context. Science is perceived as "subject to the forces of a given society" and its ethical failures as sad examples of corruption present everywhere—no more but no less reprehensible than insider stock trading or Senatorial shenanigans.[35] News reports frequently adopt this perspective when they refer to research fraud as another "white-collar crime." *Newsweek*, for example, included science in a feature on ethics scandals in the 1980s. Americans, the magazine wrote, have few illusions about their neighbors' morality and little confidence in social institutions. Over half of those surveyed by *Newsweek* believed that people were less honest than they had been ten years before, and fraud in science was cited as one more signal of that decline.[36]

Attention to scientists' standards also blends with emerging public concern about professional ethics issues overall, of the type that stimulates investigations of engineering ethics, medical ethics, and financial conflicts of interest among politicians and stockbrokers. Perhaps Americans are at a point of cultural reassessment, as they adjust their attitudes toward truth-telling and truth-tellers. It may be only natural that revisionism extends to those traditionally regarded as seekers and interpreters of truth. In museums, for example, as curators espouse new approaches to fakes and facsimiles, they attempt to fit them into historical understanding of the art rather than hide them in storerooms. Some museums, wary of trendy exhibition techniques that exploit reproductions and re-creations, now explicitly remind visitors that, whatever the smoke and mirrors employed in an exhibit, the "ultimate aim is to tell the truth."[37]

Another aspect of the measurement question involves controversial efforts to demonstrate how little "bad science" exists, compared to the vast amount of "good science." The editor of *Science*, Daniel E. Koshland, Jr., drew both applause and censure when he admitted that research fraud was "inevitable" and "unacceptable" but argued that it did not deserve much attention because it represented such a minute proportion of all research done in the United States: "We must recognize that 99.9999 percent of reports are accurate and truthful," he wrote.[38] When Koshland was rebuked for that statement, he fired back that he had not meant to imply that "scientists are ethically 99.9999% pure" but, given the millions of "bits of information" in all science journals, to note that the small amount of probable error was negligible.[39]

The question of "how much is there?" represents an inherently unanswerable question as long as it is posed in opposition to something labeled "good science." To the neighborhood that has never experienced a robbery, *one* robbery is too many. The setting is also crucial—like an embarrassing theft that takes place in a building surrounded by guards and expensive security equipment. Now that ethics offices have been established at the National Science Foundation (Office of Inspector General) and the National Institutes of Health (Office of Scientific Integrity and Office of Scientific Integrity Review), and now that these groups are not only investigating but compiling data on cases, there will eventually be comparative information on the *number* of allegations and convictions, if not also on the proportion of misconduct in research overall.[40] In the meantime, the question hangs in the air—unanswered, if not intrinsically unanswerable.

Why Pick on Scientists?

No one likes to be accused of wrongdoing, of course, and scientists are no exception. The leadership of the scientific community does not react cheerfully to either negative mass media attention or legislative scrutiny; they often complain that all scientists are being held accountable for the moral lapses of a few.

Strictly speaking, the diversity in science offers some support to this argument. There is not one "scientific community" but many overlapping ones. People with advanced education and training in the sciences, social sciences, mathematics, and engineering form distinct social and intellectual networks within their research fields, which occasionally intersect and overlap. Scientists work for all sorts of organizations, from universities to private corporations; they perform research, development, analysis, management, administration, and teaching. They may share common identities within a discipline or subdiscipline because they share informal and formal networks of communication, have comparable levels of education, use similar methods of investigation, acquisition, or authentication of data, and compete for similar rewards. They work as ornithologists, chemists, or geologists but refer to themselves generally as "scientists"; they conduct research in subspecialties of psychology, sociology, or economics and call themselves "social scientists." Each subgroup sets its own standards for research procedures, policies, communication, and evaluation, yet each group also feels part of some nebulous whole.

The rest of the world outside science distinguishes little among these subgroups. Public images of science and scientists tend to blend together the various subclassifications, blaming or praising "science" and "scientists" for knowledge or actions that are attributable to specific groups or individuals.[41] "Science museums" and "science magazines" similarly blend their content, and "science" reports in the news media follow suit. Congress maintains "science" committees (the Senate Committee on Commerce, Science, and Transportation and the House Committee on Science, Space, and Technology); in other countries, there are ministries of science; and the U.S. president appoints a "Science Adviser," head of a White House science council. In the strictest sense of membership requirements and rigid social cohesion, there may not actually be a "scientific community," but everyone acts as if there were.

Uniformity among scientists comes more from shared social norms, largely informal and uncodified, than from any common social characteristics. Science is a "community of trust" which "relies on a series of confident expectations about the conduct of others."[42] Informal rather than formal rules tend to govern research group interactions, including the process and politics of scientific communication, and there is some consistency across fields and organizations. Unwritten but identifiable standards determine "who is permitted entry" to a discipline and "how they are expected to conduct themselves once they have entered."[43] Informal rules also tend to determine who has, or who takes, the authority to investigate, expose, and punish violations of professional conduct.

In defense against what is perceived as unjustified criticism, scientists frequently point to science's reputation for accuracy. The peer review system for grant proposals and journal articles insures the reliability and integrity of research methods and conclusions, they argue. Scientists, they assert, are also trustworthy because they have so little to gain from deceit; the rare instances of fraud or fakery reflect the character flaws of individuals, not weaknesses in the research system; moreover, those who fabricate or fake data are not even "real" scientists because real scientists follow rules of honest research and do not stoop to cheating. The news sections and letters to the editor in scientific periodicals such as *Science, Nature,* and *Chemical & Engineering News* contain expressions of feelings just like these. The weekly publication *Chemical & Engineering News* contained over sixteen substantive letters in 1987, for example, most responding to a feature story the magazine published earlier that year.[44]

In the 1970s, it was more common to characterize the issue as one of personal moral conduct rather than public policy, an approach exemplified in the first case to receive wide publicity outside science, that of William T. Summerlin, a dermatologist working on skin cancer at the Sloan-Kettering Cancer Center in New York, who admitted in 1974 to a rather sensational act of misconduct.[45] Immediately prior to a meeting with the director of Sloan-Kettering, Robert Good, Summerlin used a felt-tip ink marker to darken the skin grafts on two white mice, previously allotransplanted, to make it appear that the experiments had been successful. When the deception was discovered (within hours), Summerlin claimed that he had been under considerable mental and physical strain. In his statement he referred to "mental and physical exhaustion" and "personal pressure" generated by a heavy schedule; but he also attempted to blame the "painted mouse" episode—and other misrepresentations of data eventually uncovered by an investigating committee—on "extreme pressure placed on me by the Institute director to publicize information . . . and to an unbearable clinical and experimental load which numbed my better judgement."[46] The investigating committee found that Summerlin had "grossly" misled colleagues, misrepresented the success of experiments to colleagues, supervisors, and the press, and had lied about certain aspects of his research, and they recommended dismissal. Nevertheless, their report carefully noted Summerlin's "personal qualities of warmth and enthusiasm," his "self-deception," and what it considered a "frame of mind" that gave rise to the behavior.[47]

These statements reflected sentiments within the scientific establishment which were widely accepted at the time. For example, *Science* writer Barbara J. Culliton observed that the controversy had had considerable negative effect on the other scientists involved; she took special note of the "agonizing personal ordeal" of Summerlin's mentor, Robert Good.[48] The widespread assumption, she implied, was that Good had been deceived and that he should not be ruined by the affair because *he* was a "valuable" scientist.[49] Concurrently, however, a different type of response was emerging, one less willing to dismiss misconduct as isolated, as epitomized by Culliton's comment that the "very thought of fakery threatens the powerful mystique of the purity of science." "It stirs deep and contradictory feelings of incredulity, outrage, and remorse among the entire scientific community," she continued—but the issues raised by the case, "go far beyond the question of whether one man has or has not literally falsified."[50]

Why All the Fuss?

The mass media play critical roles in raising the level of discussion on all science policy issues, but many scientists bristle at the mere hint of the negative publicity associated with fraud. Some even believe that the science press should ignore the topic altogether because the frauds are only "little mischiefs," of concern only to the scientists involved. The effort to downgrade fraud to a moral "mischief" in fact resembles attempts in the art world to treat art forgery as "less morally reprehensible than other forms of crime" because it, too, does not directly harm other people, even though it represents the same lapse in character as "cheque-forgery, larceny, [and] counterfeiting."[51]

Rejecting the significance of known cases of fraudulent research underpins another argument against discussing these issues publicly: deception may indeed have taken place, but the crimes are trivial because the work in question is scientifically unimportant. Authenticated, legitimate science is deemed superior to imitation work, even if the fraudulent work was judged superior when it was believed to be legitimate.[52] In science, as in art, "the knowledge that [a piece of work] is or is not genuine" affects its "aesthetic value."[53] If the work is later proven to be falsified or forged, then that outcome *proves* its worthlessness. As long as the number of reported cases remained "comparatively small" and involved only work deemed to be of minor importance, these arguments carried more weight.[54]

Why Not Just Let Science Police Itself?

Even scientists who admit the existence of significant amounts or types of unethical conduct may nonetheless argue for continued autonomy in vigilance, compliance, and correction because science is inherently "self-correcting" or "self-cleansing," and peer review or replication of experiments will eventually uncover false research. Certainly, the monitoring systems inherent in how research is organized—from the peer review of grant proposals, to the informal reviews of supervisors, to journal peer review—assure that, whatever the discipline, the work is always held to the toughest standards of the surrounding organization. The rallying cry for those opposed to government regulation of research has been, from the beginning, "trust the scientists to monitor their own professional conduct" because only scientists have the technical knowledge sufficient to conduct comprehensive review.[55] Such arguments perhaps seemed sensible enough in the 1970s. The Summerlin case, for example, was investigated and adjudicated privately (within the accused's institution). Self-monitoring worked. Focusing on individuals reinforced perceptions that these were problems of personal morality, not public policy. This characterization could not stand, however, in the face of what became an increasing number of episodes of a sweeping and sensational nature.

Isn't This a Problem for Only a Few Fields?

The fault always seems to lie elsewhere. The most sensational allegations in the 1980s involved biologists and, in particular, young researchers like John Darsee who had M.D.s and were working in the highly competitive, well-funded areas of biomedical research. Perhaps it was natural for physicists and chemists to blame biology, to speculate that this was not a problem for all scientists.

Fraud within biomedical research also attracted the majority of the press attention. Sociologist Allan C. Mazur believes that the bias in coverage is explained by public interest in medicine; its greater news coverage overall just heightens the impression that there are more incidents of unethical conduct among biologists.[56] The disproportionate magnitude of federal money for the biomedical sciences has been offered as another explanation for media interest—NIH is by far the greatest single federal source of money for basic research in the United States. Other commentators have pointed to the excessive tenure pressure in biology departments and medical schools.

Even biologists have considered the possibility that the seeming prevalence of cases in biology may be intrinsic in its methodological fuzziness. NIH official William Raub once pointed to biomedical research's reliance "on an observational or experimental paradigm with little theoretical base to drive it" as one explanation.[57] The competitive nature of biomedical research, and its pressure for immediate results, however, provide suspiciously convenient scapegoats for those who want to shove blame elsewhere.[58] If unethical practices can be traced to "the career structure and reward system of modern science," as part of science's competitive "publish-or-perish" game and the quest for academic tenure, sociologist Patricia Woolf observes, then blame can be directed "away from any deficits in the institution or with the [specific] work situation."[59] They can also be directed away from one's own field or institution, perhaps to those working in applied areas.

The evidence for discipline-based explanations for research fraud has weakened as more cases appear in other fields. In fact, no single part of science or scholarship has a monopoly on unethical behavior, only perhaps on particular modes or means for deception. The legal consequences of deceptive practices by authors are the same, regardless of research topic. Fabrication of data is easier in some fields than in others, but plagiarism and lying about authorship occur in all academic specialities, theoretical and experimental, clinical fieldwork and laboratory-based. The fault lies, instead, in some other factor.

Who Created the Problem?

Easy answer: the mass media. News coverage of research fraud during the 1980s, compared to how the same organizations treat political graft, was relatively tame, but the scientific community is accustomed to positive news coverage of its activities. Time and again, the mass media, those messengers of negativism, were held responsible for making research fraud into a public issue.

Science journalism still draws considerable blame for creating the controversy, even among those who admit that science must publicly reaffirm its integrity. The National Academy of Sciences report *On Being a Scientist* (1989), for example, states that attention to scientific fraud will "undermine the public's confidence in science, with potentially serious consequences" and implies that journalistic reports have "exaggerated" the frequency of fraud.[60] "Assaults on the integrity of science" are not only "unjustified" but will endanger political and economic support for research, the report concludes. As chapter 6 shows, however, the news media paid little attention to the issue in the 1970s and early 1980s; many science journalists accepted conventional explanations for what was happening and did not dig further. Eventually, an unusual, uncoordinated confluence of members of Congress and whistleblowers took advantage of changes in the political climate for science and spurred quiet debate into public controversy.

OVERSIGHT AND REGULATION: A BRIEF POLITICAL HISTORY

Scientists involved in the policy-making process tend to exploit these explanations whenever they want to derail negative publicity that could endanger federal funding. As congressional attention increased in the 1980s, inflamed by the apparent failure of major universities to investigate cases thoroughly, the fears of political harm to science seemed well-founded. The tone of crisis was, however, a natural by-product of the congressional oversight process.

Once federal research programs have been authorized and their funds appropriated, Congress still has a responsibility to oversee how well they are managed. Although a concern of Congress as a whole, oversight is carried out in practice by the committees and subcommittees of the Senate and the House of Representatives, which identify issues, determine the form and timing of investigation, and interrogate witnesses.[61] Committees exert limited but effective power either through proposed legislation (e.g., legislation that directs an agency to correct a problem or to submit a report on a specific topic) or through investigation of how (or whether) legislation has been implemented, a process that is most visible to the public during committee hearings.[62] At the extreme, congressional investigations may lead to prosecution of malfeasance; more often, the mere *threat* of restrictive legislation prompts corrective action by an agency or program.

At least through the 1970s, Congress exercised this power only very sparingly, allowing considerable autonomy to the National Science Foundation (NSF) and National Institutes of Health (NIH) in research management and they, in turn, gave their university grantees similar independence.[63] Science's ability to "self-regulate" was rarely questioned and a friendly atmosphere pervaded relations among Congress, the scientists who ran the agencies, and the research community. The House Committee on Science and Astronautics (now the Committee on Science, Space, and Technology) was established in 1958 as the first select committee designated for general science and technology oversight but it, too, tended to regard its most important purpose as insuring the health of R&D overall and promoting wise federal investment in basic research, not as excoriating agency performance and policies in its hearing rooms.[64]

Few things can appear to "sensationalize" an issue more quickly than an aggressive congressional hearing. Because a legislative inquiry is *intended* to emphasize differences of opinion, and to place the views of multiple stakeholders on the record, a tone of dissensus is set even when disagreements are minor. Media coverage forms part of the process, guaranteed to stimulate controversy, either by bringing emerging crises and alternative political perceptions to legislators' attention or by disseminating congressional debate to voters.[65] What can seem an artificial process can, however, provide a convenient tool for policy analysis; the published records of hearings impartially record the unfolding of an issue. The following analysis draws from that record, from the transcripts and committee reports, from notes and recordings made at hearings, and from conversations with participants, both those in the limelight and those standing in the wings.

The first oversight hearing on scientific research fraud took place in 1981. It is significant that it was held by the congressional committee that is, by tradition, most sympathetic to the needs of science.[66] In exploring the topic of fraud, the House Committee on Science and Technology, then chaired by Representative Albert Gore (D-Tenn.), responded to a rash of sensational cases, most notably the John Darsee case, that science's strongest proponents simply could not ignore.

At the hearing table on 31 March sat senior officials of NIH and other government advisors, expressing views widespread within the scientific community at the time. Philip Handler, president of the National Academy of Sciences, called the problem "grossly exaggerated," and he was supported in this characterization by NIH officials, who denied that fraud was widespread.[67] But the members of Congress seemed unwilling to accept that dismissal or to focus on philosophical issues; instead, they seemed concerned about how NIH was handling its own investigations. Why, one congressman asked, was NIH continuing to fund scientists who had been formally accused of wrongdoing? When NIH representatives explained that investigations took a long time and they did not want to appear to be "blacklisting" a scientist who might later be acquitted of charges, the members of the Committee seemed incredulous. When the officials further testified that no scientist was then being disqualified from eligibility from NIH funding *even after admission of guilt*, several congressmen testily pointed out that, as they understood the law, the "presumption of innocence" rule applied only to punishment, *not* to entitlement to federal funding.

This hearing and others that followed in the 1980s contained an unmistakable message from Congress to the leaders in the scientific community, a message that was politely, emphatically, and repeatedly stated but seemingly ignored. For reasons related partially to the lack of an internal consensus on the issue (e.g., whether fraud was a problem, how much of one, or who should fix it), the scientific community appeared to dismiss the political warning signs and to be unconcerned about punishing even admitted acts of fraud and plagiarism. Delays in the regulatory rule-making process within the federal science agencies further added to congressional concern that neither the scientists, the universities, nor the middle-level government officials who monitored them really wanted these cases investigated, much less resolved. Although the NIH began in 1981 to develop formal policies on how its institutes and its grantee institutions should conduct investigations, it did not publish agency guidelines until five years later and its institutional guidelines, not until three years after that.[68] Many colleges and universities also waited until 1989 or 1990 to begin developing comprehensive internal procedures (in response to the federal requirement that they do so); so the academic world, too, appeared complacent.

By the mid-1980s, the issue had begun to escalate to the level of public controversy, largely because of this resistance and the anger of critics who demanded action. At the 1983 annual meeting of the American Association for the Advancement of Science (AAAS), a session on "Fraud and Dishonesty in Science" brought together representatives of various perspectives on the issue, ostensibly to discuss *Betrayers of the Truth*, a book by science journalists William Broad and Nicholas Wade, which had been attracting fire from scientists for several months.[69]

The book charged that the cases of scientific fraud uncovered so far represented only "the tip of the iceberg," that much more ethical misconduct existed in science than scientists were admitting, and that, given the substantial federal investment in research, any failure to acknowledge this situation was a matter of public policy. The AAAS session featured a number of speakers but one especially hostile debate between Wade and biologist Norton Zinder stands out because it highlighted well the differences between the two dominant ideological perspectives: those who believed the problem to be simply one of a few "rotten apples" (Zinder) and those who regarded science as a "rotten barrel" (Broad and Wade).[70]

In retrospect, the positions of Zinder and the two journalists do not seem so incompatible. Zinder argued that cheating was a human trait, that it was not confined to science, and that the peer review system had never been designed to detect fraud. Broad and Wade argued that not all fraud was detected, that there was more fraud than was being acknowledged, that "careerism" was affecting ethics in science, and that the peer review system and other presumed checks in science could and did fail.[71] All six of these assertions were true, of course, but neither side was willing to concede common ground.

The assertions of *Betrayers of the Truth* were also not all that radical. The journalists had simply focused on a logical inconsistency in how scientists had tried to frame the political debate: the sensational cases of fraud had not been uncovered by traditional peer review checks and yet those mechanisms were continually proposed in congressional testimony or in statements to the media as the means of insuring that fraud was rare, or as proof for the reliability of scientific expertise and the accountability of scientists. Through the 1980s, as the inconsistencies became more obvious, the interpretation advanced by Broad and Wade gained legitimacy. Eventually, the debate began to concentrate less on the so-called pathology or deviance of individuals (the "rotten apples" theory) and more on the public policy implications of systemic corruption and failures to investigate it.

In attempting to understand the history of political and institutional action on fraud, it is helpful also to recall the overall political atmosphere of the 1980s, because conservative emphasis on de-regulation and the resistance to federal interference in research undoubtedly influenced the pace of response from the agencies. The Reagan administration had a strong commitment to R&D, especially as related to national defense and international competitiveness; increased regulation of the research process was perceived as a potential threat to those goals. Yet, equally strong social and political forces pulled *toward* regulation. Attention focused on fiscal accountability and the quality of research management in the universities. Organizational changes within Congress, which had resulted in increased emphasis on legislative oversight, helped to turn legislative attention to research problems.[72] Moreover, scientists no longer served as the only advisors on science policy issues. Voices other than those of scientists commanded attention; when congressional committees discussed science policy, they heard testimony from sociologists and historians of science, policy analysts, and economists, as well as from Nobel laureates and federal research managers. As Representative George Brown (D-Calif.) of the House Committee on Science, Space, and Technology observed then, Congress was moving away from "the era of unthinking acceptance of the wisdom of anyone with scientific credentials."[73]

Federal agencies, universities, and professional societies considered how to remedy gaps in their policies and procedures, but the number and extent of cases appearing in the press continued to give an unmistakable impression of widespread misconduct. In October 1986, for example, a University of California-San Diego committee announced the results of its extensive internal investigation of a rising young scientist on its faculty, declaring almost half of his published work to be either "fraudulent" or "questionable" (a case detailed in chapter 4). So, by 1988, many influential members of Congress had begun to lose patience with the pace of investigations and the development of institutional policies. Some type of federal action had been proposed periodically, by various Congressional committees, for over seven years, but little had taken place. Within the NIH, administrators stated privately that the notion of fraud in the biomedical sciences was "a maliciously constructed illusion," "a trivial matter gone berserk."[74] In response, Congress began to use its oversight powers to shake things up.

In one particular month, April 1988, a series of events took place that demonstrated the issue's political volatility. On 11 April, at a hearing of the House Committee on Government Operations, scientists who had "blown the whistle" on fraudulent research described how they had been vilified by colleagues and senior administrators and how the investigations prompted by their allegations had been repeatedly delayed at all institutional levels.[75] On the next day, a hearing of the House Committee on Energy and Commerce's Subcommittee on Oversight and Investigations featured two young scientists who testified under subpoena. They described how their questions about experimental data reported in a journal article had been rebuffed by administrators and colleagues at MIT.[76] Because one of the coauthors of that article was the Nobel laureate biologist David Baltimore, the hearings made front page news. Committee Chairman John D. Dingell (D-Mich.) told NIH officials at one hearing that the committee intended to continue its investigations with vigor: "We are going to be visiting again about these matters," he promised.[77] Legislation introduced in the House of Representatives (H.R. 2664) attempted to criminalize the knowing misreporting or misrepresentation of scientific data to the federal government.[78] And, on 15 April 1988, a research psychologist, Stephen J. Breuning, first accused of misconduct in 1983, was indicted on federal charges of having falsified research project reports to the National Institute of Mental Health (NIMH).[79]

The Breuning case demonstrated once and for all that fakery and fabrication could not be ignored in the vain hope that, somewhere down the line, science would "self-correct" and no harm would be done. When later in 1988 Breuning pled guilty and was sentenced to community service and a substantial fine, the facts of the case showed more than "intellectual mischiefs." Breuning's research data had, in fact, been widely reported and accepted by practitioners; from 1980–1983, his published papers represented at least one-third of all scientific articles on the topic; a citation analysis of his publications concludes that, from 1981 to 1985, his "impact on the literature was meaningful."[80] Moreover, the results of his drug experiments (many of which had, in fact, never been conducted at all) had been used to justify the treatment plans for hundreds of hyperactive children. There is no evidence that any patient was harmed. Nevertheless, the Breuning case emphasized a terrible possibility: that faked or misrepresented scientific data *could* actually persist in the literature long enough to be accepted as accurate and to risk possible harm to innocent parties.[81]

The hunt was on. The "illusion" that the universities and the leaders of the scientific community had things under control had been brutally dispelled. Through the oversight process, congressional committees began to demonstrate forcefully that they did not consider these matters to be "trivial."[82]

Since that time, congressional hearings and the eventual implementation of agency regulations have prompted energetic discussion of research ethics, reports from committees of scientific groups, and the development of formal guidelines for how institutions must investigate allegations of research fraud.[83] An annual round of oversight hearings on the issue appears to be commonplace. In 1989 and 1990, hearings included testimony by document experts from the Secret Service, who had been asked by a congressional subcommittee to scrutinize forensic evidence in an ongoing NIH investigation; in 1991, they examined the NIH ethics investigation process itself. The atmosphere in these hearings has been charged. The rooms are packed. Whistleblowers and the supporters of the accused jostle for seats with a blossoming number of professional staff members conducting studies at federal agencies and scientific academies. Members of the press wear guarded but skeptical expressions. And everyone exchanges knowing glances, trying pretentiously to look like insiders. During the questioning, discussion increasingly moves along nonpartisan lines as the issue intertwines with other policy issues related to the state of U.S. science in general. And most important, the climate is far less warm and conciliatory than it was at the 1981 Gore hearings.

A simple conclusion that one can draw from the changes is that scientists apparently do not learn political lessons as easily as they learn new lab techniques. As late as 1989, at a small meeting in Woods Hole, Massachusetts, between congressional staff members, senior scientists, and university officials, there was still considerable argument about whether the federal government, and especially the Congress, "should" be involved in this issue. Observers at the meeting later described how, after many heated exchanges, the scientists just suddenly seemed to awake "as if from a deep sleep" and "to comprehend how Congress views the problem." Some of the university officials present claimed that, until those discussions, they had been "unaware" of the intensity of congressional attitudes toward research fraud.[84] That was a lesson late in learning. Even though support for funding scientific research remains high in the 1990s, the rhetorical signals indicate increased political skepticism and decreased autonomy on the horizon. Without question, congressional committees will no longer regard research fraud as either "not a problem" or "best left to the scientists to fix," as the witnesses in the 1981 hearing had pled.

NOT JUST "NEW POLITICS," BUT "NEW ECONOMICS"

Why did many scientists react so hostilely to public attention to these issues? Why does the discussion of scientific fraud continue to generate more heat than light? Edgar Allen Poe provided one answer—a natural human repulsion. Scientists share that visceral reaction and the intensity of it can make it difficult to remain quiet. Another answer lies in the lack of consensus on either a root cause or a best solution, which leaves participants feeling powerless. But a third factor helps to explain why the controversy arises now, when science is so visibly successful: the changing political and economic environment for research. Competition is keen, for resources and for power. Scientific knowledge is a major new commodity.

The central themes of two seminal books in science policy illustrate this change well. In 1967, when Daniel S. Greenberg published *The Politics of Pure Science*, he broke new ground by arguing that, from the post-1940s era (when "science was an orphan") to the late 1960s, scientists were indisputably engaged in "politics" whenever they advised the government, even if their motives may have been pure.[85] They could not claim to be *a*political. In 1984, following the lead of Greenberg's last chapter, another science journalist, David Dickson, considered what he called *The New Politics of Science* and described a community that had become even more politically astute and economically significant, entwined in the commercialization of research, the international competitiveness of U.S. technology-based companies, and the interconnections of world communications.[86] Although he considered whether science could ever be "democratized," in the sense of responsiveness to the public will, Dickson implied that the answer was no. The interests of scientists, government, and the national economy (and the interconnections between them) were entrenched.

What shines through the writings of Greenberg, Dickson, and similar analysts of contemporary science policy is not just the "new politics" of science but also its "new economics." The controversy over scientific fraud and misconduct, although initially framed as a philosophical or psychological problem, as a disagreement over standards of morality, exemplifies this shift for it now derives much of its political energy from assumptions about the value of scientific knowledge and expertise. Scientific ideas seem to be worth stealing. Like literary plagiarism, which takes on meaning when authorship achieves value, scientific fraud gains importance from its sense of economic worth. As perceptions of the commercial value of knowledge generated through *basic* research have grown, the perceived profit from the theft of research ideas, from fabrication of data, or from the deception committed in order to attract or ensure funding have increased proportionately. In even the "smallest" robbery, it is the perception of gain, the dream of profit, that motivates risk-taking. A booming economic climate provides new opportunities. Michael Levi writes, in his book on "white-collar" fraud, that "an aspect of crime too often neglected by criminologists is that as other forms of social and economic organization change, the forms of criminality change . . . no computer fraud without computers, and no credit-card fraud without credit cards."[87] Levi pleads for analyses that integrate understanding of "potential offender," "potential victim conduct" (e.g., in the case of research fraud, how organizations operate), and the social and economic context. Indeed, something other than the mere thrill of seeing one's name in print must have motivated these bright accomplished scientists to risk theft of ideas or risk fakery or forgery of data. If we care about the integrity of the research process and the knowledge base, we cannot simply shake our heads and turn away in disgust. We must take an honest look.

Economic motives undoubtedly seem distasteful to those who believe that science is driven by an intellectual engine. Yet, scientific information, insights, and data do represent commodities apparently ever-more precious or desirable to ever-greater numbers of people and organizations. Billion-dollar corporations invest in fundamental research (i.e., not-yet-applicable research) on superconductivity and biotechnology, anticipating profit. Universities see patents as income. In such a business climate, the most arcane piece of information may appear valuable when possessed by a competitor. This changed perception of value affects both formal and informal communication among scientists. Laboratory colleagues who ten years ago talked freely in the corridors now say that they describe their ongoing research only in generalities or euphemisms, lest they inadvertently reveal marketable knowledge to competitors. First publication in a journal determines not just priority and prestige but can also help to claim lucrative intellectual property rights.

At the same time that perceptions of value are changing, there have been major shifts in government funding of and regulatory controls on scientific research practices and communications. Before the 1940s, U.S. economic support for scientific research—especially basic research—was far from lavish and came more from industry, universities, and private foundations than from the federal government. Influential leaders in the scientific community even considered federal research grants to private universities to be inappropriate and possibly unconstitutional, an attitude linked to their desire to preserve autonomy in how research was conducted and communicated. Because science had not yet "proven" its utility through wartime service, there seemed little for the nation to gain from subsidy of even the most promising basic research. Although exploration of fundamental questions was widely regarded as "worthwhile," science was considered to be valuable only when it was applied, that is, once it became technology. Aside from this vague potential, the insights, theories, or processes of basic researchers appeared to have little intrinsic commercial value, thus ideas were freely shared among colleagues.

Increased federal support for research, and its subsequent success, influenced political perceptions and economic realities. As late as 1940, the federal government accounted for only one-fifth of the $345 million annually spent for scientific R&D in the United States. The Manhattan Project and other war-related research dramatically altered those proportions, just as they also altered the relationship of basic science to government. Science proved to be a useful tool for government, and few politicians or scientists suggested a return to the pre-war "hands-off" relationship.

As they bargained for new managerial models and higher levels of federal support, scientists attempted, with some success, to reserve old-style managerial autonomy: *scientists* would stock the expert review committees allocating government funds, thus deciding how (and which) research should be done, and scientists would monitor their own professional conduct. This model placed considerable trust in researchers' productivity, reliability, integrity, and accountability. It also assumed that scientists would utilize government funding prudently and wisely and, in return, the products of their research would stimulate the economy and commercial innovation. It was a relationship that functioned well as long as scientists appeared to be models of virtue, and as long as the products of science flowed without interruption into the national economy. When cases of unethical behavior came to light, they were at first tolerated because they were believed to be rare or insignificant.

The issue of fraud in the conduct and communication of science—as a policy issue—therefore revolves ultimately around the concepts of *autonomy* and *accountability*, two aspects of science's political presence which are currently set in tension. "Scientific freedom," exploited as a metaphor for the personal and intellectual independence of scientists, is a red herring in this debate, for individual autonomy at the laboratory level is not in dispute. Instead, concerns center on "collective" (or institutional) autonomy in the management of federal funding. The "price" for funding—in terms of the independence of science as a collective activity—became apparent in the 1970s when those who opposed particular aspects of science had the political tools to extract cooperation, and when the federal government imposed new restrictions on scientific communication related to national security. Regulation of experimentation on human subjects or animals, even the toxic waste disposal practices of laboratories, all became issues of public concern, as ordinary citizens energetically engaged in local political actions that interrupted the work of research labs within their communities.

In the 1980s, regulations that applied to academic institutions receiving federal funding had the unintended effect of removing a layer of protective autonomy for science. In all government-funded research, neither the precise choice of topic nor the approach had ever really been left to individual discretion; instead, agencies issued requests and set the requirements to which projects must conform.[88] What *had* remained through the 1980s was collective autonomy—in that the scientific community shaped policy making at the national level, influenced the broad directions of research programs, and influenced the acceptability of certain research topics and methods. As the community itself has become increasingly fragmented in its political responses to funding and regulatory crises, then that power has also diminished.[89]

Accountability is the other watchword. How scientists conduct themselves within the privacy of their laboratories has become a matter of public policy, if the research is federally funded. Lying in a grant proposal, lying in a report to the NSF, stealing ideas from someone else's NIH proposal, or reporting falsified results from a government research project—phrased thus, what some term "little mischiefs" appear as appropriate matters for congressional attention.

Americans still tend to perceive scientists as truth-seekers *and* as truth-tellers, but political trust has been shaken. The peer review system that validated scientific knowledge seems fallible. And the prevalent questions themselves show the depth of concern. Do these cases indicate widespread lack of accountability among authors, referees, or editors? If dishonesty is rampant in science and if the peer review system can and does fail to detect fraud, then how reliable can the scientific knowledge base be? And more important, can we trust this knowledge sufficiently to use it as the basis for other public policy decisions? Questions such as these, especially when hurled about in a congressional hearing, have transformed the issue from concern with individual deviance to explicit preoccupation with federal science policy. Criminal indictment represents a far different process for seeking resolution of wrongdoing than the behind-the-scenes methods of the *Lancet* or Dr. Blackman of Cincinnati, with far different implications for how the norms of conduct may be defined—and by whom.

As long as unethical research conduct was prohibited, detected, and published in the close confines of academe, then crispness of definitions probably seemed less important. All participants spoke the same language and the appropriate procedures were part of the unwritten culture of the group. Once disputes moved into the public arena and the number of parties to a dispute increased, the ambiguity inherent in common allegations became apparent and debilitating. Without precise definition of *whose* rules or standards have been violated (much less *which*), investigations become stalled and resolution impossible. Blackman could point to a specific inappropriate act—the word-for-word translation of text without citation to original authorship; modern disputes include such diverse issues as who should be listed among an article's coauthors and what constitutes theft of ideas in a cooperative, multiple-organization research project.

A single hour of a congressional hearing on the topic reveals clearly that no common glossary of standards, neither high nor ordinary, exists. Moreover, as the next chapter describes, the definition of what constitutes a violation of standards for scientific publishing depends most on the initial perspective—that of science, publishing, or public policy.

2   Classifying Violations

A rule has been broken, a wrong has occurred, even though the rule
may never have been written or discussed. Nevertheless, the person
wronged assumes that both parties have subscribed to the rule.
                    —Richard H. Blum[1]

Human beings are rarely as consistent in their attitudes or behavior as
they pretend to be. Perhaps out of fear of Emerson's "hobgoblins" or
perhaps from centuries of evolutionary adaptability, a person may blatantly violate the law in one situation and yet be meticulously well behaved
in another. In the context of scientific and scholarly communication,
professional misconduct cases have included multiple violations of legal,
social, and moral norms by individuals whose lives and careers appear
otherwise above reproach. An author may be scrupulously fair in assigning
proper credit in one article, yet plagiarize repeatedly in another. A scientist
may fabricate data for one set of experiments yet carefully conduct real
research in every other. Such paradoxes reinforce the need to establish
"cultural neutrality" in determining where misconduct has occurred, to
allow for variations in standards and practices, and yet to reduce the
tendency to indulge in "ethical relativism."[2]

When disputes arise over the definitions of research misconduct, such
as what actions are being questioned and why, the disagreements do not
represent mere rhetorical quibbles. They frequently reflect important
differences among the participants' value systems and codes of professional
conduct as well as ambiguities in institutional procedures and policies. For
disputes over research communication, clear lines can be drawn between
illegal acts (prohibited by federal, state, or local laws, or by government
or formal institutional regulations), *improper professional conduct* (what
publishing policies dictate as proper behavior for authors, referees, or
editors, and the norms for professional conduct which may be articulated
in association codes or during graduate training), and *immorality* (behavior
that violates a society's moral standards).[3] Unfortunately, the controver-

sies over research fraud, especially those that have arisen within academic institutions, have too often been characterized by labeling an accused's behavior as "right" or "wrong" without identifying what (or whose) rules, codes, or norms have been violated. This chapter lists behaviors by authors, referees, and editors which are widely regarded as violating accepted standards in the publishing process. I then discuss how the definitions may be influenced by the diverging perspectives of the law, the scientific community, and the world of publishing.

Whether an act violates legal, social, or moral standards eventually determines the extent and likelihood of institutional response. It dictates who (if anyone) from what organization is responsible for investigating; it influences whether participants believe a formal inquiry is warranted; it proscribes how, when, and in what forum any inquiry will take place; and it affects the objectivity and fairness of investigation and resolution. For example, in their article on John Darsee's coauthors, Walter Stewart and Ned Feder condemned the practice of "honorary" authorship. That practice violated no state or federal law in place at the time; it did violate unwritten expectations at the journals in which Darsee published. And yet, the awarding of honorary authorship seemed at the time to be standard practice in biomedical science, such that it drew little open condemnation even from those scientists who privately rejected the practice. A second example, the plagiarism of text from another person's article, violates federal copyright law as well as principles of good scholarship if the plagiarism or the plagiarized text is disseminated publicly (e.g., "published"). The first step in analyzing any accusation or suspicion of publishing misconduct is therefore to sort out, according to every participant and institution involved, which actions may have violated which applicable standards for conduct. Only once it is decided whether standards are governed by law, regulation, contract, custom, or social morality can one begin to determine what should be done and who should do it.[4]

Making such determination can be doubly difficult if the work is interdisciplinary, interinstitutional, or international, for few of these standards apply across all disciplinary, organizational, and national boundaries. The rich diversity of the scientific community—an asset in most circumstances—hampers attempts to set uniform standards for professional conduct. Whenever scholars judge their colleagues' actions, they also tend to impose criteria peculiar to their own discipline, even though correct practice in one field may be unacceptable or inexcusable in another. This variation can impede the work of investigating committees and compounds the difficulty of developing standards applicable to all types of research communications and acceptable to the nonscientific interests caught up in disputes.

Another critical circumstance involves the conflicts of interest related to the different professional allegiances and obligations of participants. A journal editor's education, training, and research interests tend to coincide with those of the journal's readers; but that same editor must also impose the publisher's standards and legal rights in any dispute, possibly forcing him or her to contradict mainstream opinion in science. As chapter 3 discusses, publishers worry about economics and circulation trends, concerns that can seem irrelevant to university scholars who focus only on a journal's content. Other so-called outsiders involved in a dispute, such as lawyers or members of Congress, bring still other sets of values, expectations, and biases to the negotiating table.

In addition to attitudes related to their disciplinary affiliation or administrative role, people often rank misconduct on a scale of significance. Is this a major violation that must be acted upon immediately, or a minor one best handled cautiously and quietly? Marcia Angell, deputy editor of the *New England Journal of Medicine*, once proposed a medical analogy for classifying cases of scientific fraud into "benign" (e.g., loose authorship or repetitive publication), "offensive" (e.g., selective presentation of data), and "malignant" (e.g., plagiarism).[5] Other commentators believe that only the violation of written rules or specific legal obligations (as in the unauthorized publication of copyrighted material) deserves the label of "fraud." To others, *real* scientific fraud only occurs when a researcher distorts scientific data for unethical motives: "Fraud occurs when the conclusions of scientific research are blatantly at variance with the known facts; when lying is plausible, it is a misdemeanor."[6]

Definitions tend to change with time. Harry M. Paull, in *Literary Ethics*, writes, "It is commonplace in ethics that practices once deemed innocent became gradually to be regarded as crimes as civilization advances . . . the standard of morality changes with the ages."[7] Curiously, both this view and its opposite may be said to have dominated in science from time to time: practices once condemned as inappropriate or unethical have been regarded as acceptable behavior in certain circumstances—and vice versa.

These concerns extend beyond mere inconvenience, however, to questions of justice. Imprecise definition endangers an investigation's fairness. When psychologist Anthony N. Doob later discussed the deliberations of an investigating committee on which he had served, he said that the group had been seriously hampered by the ambiguous nature of the original allegations, made by people with quite divergent opinions about what rules had been breached:

One of the few nice things about being charged with a criminal offense is that the charge is clear: a specific person is charged with having committed a specific offense at a more-or-less specific time. The prosecution, then, has the responsibility of trying to prove beyond a reasonable doubt that the offense took place as charged, and the accused gets a chance, if he wishes, to mount a defence.[8]

In this case, the committee first had to decide *what* to investigate (e.g., research practices or research conclusions?). One accuser had used the term "fraud" in statements to the *New York Times*, whereas another had stated on the CBS program "60 Minutes" that she cared only whether the recommended treatment had worked as described, not whether there was "fraud" during the research process.[9] Because the committee released only limited information about the facts of the case, Doob observed, innuendo and gossip led many people in the field to persist in conclusions about the researchers' guilt which were at odds with the committee's ruling. The climate of suspicion led to equivocal reception of the committee's report, and the investigation seriously harmed the accused researchers' careers even though the committee had rejected the allegations. Such stories reinforce the need for clearly defining the applicable standards and for establishing uniform procedures *before investigations begin*. The authority of rules are what grant an investigating committee the power to render judgment, but freeing an inquiry from ambiguity represents the first step toward a just resolution.

DETERMINING HARM AND INTENT

Although the legal concept of fraud may set too high a standard of proof, the use of such a test can help to separate intentional, malicious deception from its more benign cousins: behavior intended to deceive but not to harm (e.g., hoaxes, parodies, and jokes) and actions that may cause harm but do so unintentionally (e.g., mistakes and accidental errors). The notion of "intent" can also serve as a starting point for analysis.

The world of art provides a number of helpful analogies. In art, "whether to call [a] work fakery, fraud, reproduction, facsimile, copy, or restoration depends on an intricate structure of understanding between maker, seller, purchaser, and later generations of sellers and purchasers." "What it all comes down to is intent," museum curator Jonathan Fairbanks concludes.[10] The "essence" of a forgery, for example, lies in "the deceitful intention of the forger," not in the work itself.[11] Paul Jett of the Smithsonian's Sackler and Freer Galleries has called this characteristic the "merit" of fake art or antiquities—the *object* is not the fake, the maker is.

Intention thus becomes "all-important" in determining what is a forgery and what is genuine.[12]

The presence of labeling provides one clue. Museums, for example, frequently use facsimiles in exhibits if curators are convinced that long-term exposure to light (or humidity) might damage an irreplaceable object. Most museums now reproduce (or allow others to reproduce) some of their holdings; these reproductions are then offered for sale in gift shops and catalogs. Properly labeled, of course, neither facsimiles nor reproductions defraud either visitors or purchasers. "Trouble occurs," Fairbanks points out, "when the attempt is to fool rather than to re-create," or when information vital to understanding or evaluation is lost or unavailable (as when a label has been deliberately removed).[13]

Similar distinctions illuminate the boundaries of ethical misconduct in science publishing: did the accused intend to deceive, and with what purpose? In one sense, fraud in the communication of purportedly original research resembles fraud in financial accounting, when an accountant, bookkeeper, or manager attempts to show "a false financial condition."[14] Authors who describe imaginary or fabricated data, who misrepresent research success or procedures, or who, through plagiarism, proffer someone else's ideas as their own, present a false intellectual image, usually intended to make them look more productive, more successful, or (as editor Stephen Lock has observed) "more original and creative" than they actually are.[15] Deception, of course, draws on all sorts of motives. Power, greed, fear, jealousy, career advancement, and favors or protection from the powerful have all served as the ends for which deception has been the means. In commercial fraud, the bunko artist or embezzler usually seeks quick and easy money.[16] In research fraud, the goal is more likely to be fame, promotion, or professional advancement than ready cash.

In this book, I use such terms as "misconduct" or "wrongdoing" to distinguish *actions intended to deceive for some unethical motive* from actions that may have unintentionally deceived (what we generally call "honest mistakes") or may have deceived for an acceptable motive (such as literary hoax or satire).[17] Whatever the specific motive or means, the perpetrator must have an unambiguous *intention* to deceive, to make a thing appear to be what it is not.[18] That definition attempts to be sufficiently lenient to allow for solitary instances of inadvertent plagiarism, perhaps caused by sloppy research practices or by errors in typesetting, yet be sufficiently strict to encompass obvious patterns of plagiarisms or fabrications committed with unambiguous (or provable) intent to deceive.

Comparison with the world of art is instructive here, too, for art admits a spectrum of "deception," at one end desirable and acceptable for aesthetic reasons and at the other end rejected as inappropriate and unethical. Art historians reserve the most serious term "forgery" as "a moral or legal normative concept" that "refers to an object which, if not necessarily aesthetically inferior, is always morally offensive" because it involves deception. The most "notorious and disconcerting" forgeries are those of "deliberate . . . conscious deception," when someone "has passed off a work of art as being something which it is not."[19] At the other (i.e., acceptable) end are those works that do not attempt to conceal their deception. The ingenious glassblowers who made the glass flowers in Harvard University's Peabody Museum, which at first glance appear astonishingly real, did not intend them as "forgeries of flowers" but as copies for botanical study. The student of Chinese bronzes who attempts to duplicate an antiquity yet signs it with his own initials creates a *copy* not a true forgery.[20] To determine art forgery, scholars look at the interplay of three things: the object, its producer, and its effect. Has the artist attempted to hide the fakery from the audience, to pretend that the object is real? And, if so, for how long? Is deception part of the aesthetic experience (as in trompe-l'oeil) or is it simply a cruel hoax? The joy in such works as William M. Hannett's painting "After the Hunt" or the seventeenth and eighteenth century "letter rack" paintings (*quodlibet* or *vide-poches*) comes from the viewer's surprise that the objects are not three-dimensional, as they appear from afar, but instead are artful constructions.[21] The game is illusion; the deception, that of manipulating spatial perceptions. Similar aspects characterize the work of magicians; they must have sufficient technical expertise to accomplish a trick and yet somehow let the audience *know* that "magic" is afoot. Applause acknowledges the magician's skill at temporary deception.

The term "scientific fraud" also could be assumed to imply trick or deceit, but this trick is created for unethical motives, not for education, entertainment, or enlightenment. Therein lies the most difficult problem for analyst or investigator. Attorney William A. Thomas emphasizes that, in law, "intent is always difficult to prove and . . . requires careful examination of conflicting . . . evidence."[22] In law, five elements constitute the offense of fraud, each essential for a conviction. There must be: "(1) a false representation, usually of a factual nature, (2) by someone who knew or should have known that it was false, (3) with the intent that someone will rely on it, and (4) someone did rely on it, and (5) as a result suffered a loss (i.e., incurred damages)."[23] Magicians and artists of illusion commit "fraud" only up to a point; they intend enjoyment and entertainment, not harm. Without the deliberate revelation at the end that magic has taken place, a magician's performance would have a quite different effect, of course. The assistant must emerge from the magician's cabinet intact, not dismembered, so the audience is amazed but reassured.

Hoaxes, Parodies, and Jokes

Hoaxes, parodies, and jokes are hallmarks of literary and scientific worlds—welcome interruptions in all-too-serious academic pursuits—yet the commission of an elaborate scientific hoax could easily be mistaken for more serious misbehavior, especially if it involves the fabrication of data or artifacts. True hoaxes and jokes have innocent goals: to have a good laugh, to "put one over on" one's peers, or (as in the case of hoaxes intended to fool literary critics) to embarrass one's "betters." Gerald O'Connor describes the hoax as a natural part of popular culture, common across societies, social groups, and occupations, and he classifies them into two types.[24] In a "satiric hoax," the practical joker defrauds victims for the purpose of ridicule; once the victims are snared, then the hoax is revealed and everyone laughs. The "nonsatiric hoax," however, is intended to benefit the hoaxer.[25] Literary hoaxes, which Harry Paull calls "misdemeanors" when they are directed at such goals as parody, are of the first type.[26] Literary scholars may be fooled, embarrassed, or lampooned, and their reputations may even be damaged, but the hoaxer derives little material benefit. The means, not the ends, draw our attention. Satirical publications in the sciences and social sciences (e.g., *The Wormrunners' Digest* and *The Journal of Irreproducible Results*), where the joke is apparent to the knowledgeable but not necessarily to the novice, fall into this category as well. Other hoaxes, such as health hoaxes and "get-rich-quick" schemes, use deception as a means to an end, most notably the acquisition of money or fame, and they may violate laws against false representation or false advertising.[27]

In checking allegations of scientific fraud, the possibility of hoax should be eliminated early, lest an innocent person be caught in the potentially damaging process of a formal investigation. Although deception is essential for their success, most benign scientific hoaxes seek to demonstrate (sometimes with an immature, exhibitionist flair) the prodigious intellectual or technical skills of the hoaxer; he or she may anticipate mild embarrassment or surprise among the victims but does not really wish them harm. Intellectual hoaxes, Dennis Rosen points out, may even provide "welcome . . . relief from the tedium of most papers published in the learned journals."[28] He believes that most scientific hoaxes "give themselves away by an excess of schoolboy humor," whereas deliberate frauds tend to "merge undetected" in the literature of their fields.

The possibility of hoax must also be eliminated because the elaborate technical joke approaches an art form in many science and engineering schools and plays an important role in the pedagogy. At places like the Massachusetts Institute of Technology (MIT), for example, administrators have long tolerated ingenious undergraduate pranks as long as neither property nor public health and safety are endangered. The student who returns from classes to find his dorm room turned into a putting green, complete with natural grass, cup, flag, and golf clubs, is expected to marvel at the techniques and speed with which the prank was accomplished and to dismiss any inconvenience it may cause.[29] The other type of prank may be undirected—as novelist Michael Innes notes: the best practical jokes tend to be disinterested, without a specific individual in mind as a victim and with a minimal "element of malice."[30] Universities in the United States are required by federal regulation to investigate every allegation or discovery of scientific misconduct and they now face the dilemma of distinguishing quickly between acceptable actions like student pranks and their unacceptable relatives, while not inhibiting good-natured fun.

The hoax played on an unsuspecting scientist may not always be benign, of course. Perhaps the most unusual, extensive hoax recorded in the history of science was that played on Dr. Johann Bartholomew Adam Beringer (1667–1740), a pompous physician who "plunged himself into the study of oryctics, or 'things dug from the earth'."[31] Attracted to certain types of stones with "natural figures" on their surfaces (e.g., three-dimensional reliefs of frogs, snails, scallops, and the like), Beringer began to assemble a cabinet of these curiosities and hired local boys to dig for them. Beringer then published an extensive scientific treatise (*Lithographiae Wirceburgensis*, 1726) about the discoveries. Unfortunately, the stones were neither fossils nor ancient crafts. They had been fabricated and planted in the digging site at the instigation of jealous colleagues. When the hoax was revealed, the effect on Beringer was disastrous. "The prank . . . had succeeded too well," two historians have observed, for Beringer's attempt to buy back all the copies of his book ruined him financially; he died in "chagrin and mortification."[32] This anecdote conceals a serious issue: how to treat the scientist who fabricates data to fool a colleague who then unwittingly publishes or reports it?

When a hoaxer uses fake scientific data to deceive an unsuspecting public rather than friends or colleagues, and especially when money is involved, then fraud in the legal sense may have occurred. Alexander Klein's *Grand Deception* (1955) lists a number of famous technical hoaxes committed for the sake of profit, from salted diamond mines to pseudomedical miracle cures; Klein portrayed deception, and its partners naiveté and greed, as universal impulses.[33] Frauds such as the effort by nineteenth-century inventor John Worrell Keely to promote his "minimum-input" motor have drawn on scientistic language to sell their impossible dreams; science and engineering related frauds may be found in many parts of life today.[34] Recent scandals over the deliberate forgery of research data in the pharmaceutical industry represent less amusing instances of scientific fraud for profit. The impetus for these scams lies in financial rather than intellectual gain and they represent the moral opposite of the student prank. Without question, lines should be drawn to distinguish among these various types of hoaxes, lest the jester, however thoughtless, be prosecuted and punished on an equal basis with those who deliberately and systematically attempt to defraud for personal or financial gain.

Accidents and Mistakes

Another type of ethically ambiguous conduct falls under the heading of unconscious "self-deception" and involves inadvertent errors or "honest" mistakes (as distinguished from the legitimate statistical concept of error). Considerable controversy surrounds this type of conduct when the mistakes have been published in peer-reviewed journal articles or, as in one recent "discovery" in astronomy, when the erroneous results have received international press attention. In the astronomy case, researchers described a signal that they initially believed came from a superdense star; a year later, they discovered that the signal had actually emanated from the electromagnetic interference of an old video monitor used to guide their telescope.[35] These accidents seem ridiculous in retrospect, but as the editor of *Nature*, John Maddox, emphasizes:

> Journal editors, if they are honest with themselves, will acknowledge that much, perhaps most, of what they publish will turn out to be incorrect. Neither we nor our contributors can be blamed for that, for that is the nature of science.[36]

Each new piece of information pushes scientists closer to the truth, although sometimes that may take them farther from an earlier interpretation they sincerely believed to be true.

Excluding error or accident from definitions of fraud and misconduct should not be construed as condoning sloppy practices in research, but the appropriate arena for prevention and investigation of accidental errors is the laboratory. Unless the errors appear to be systematic, part of a pattern of sloppy practices, they must be treated differently. In *False Prophets*, Alexander Kohn calls scientific errors "forms of aberration in science" that are "unintentional" and "based on good faith."[37] He argues that sloppiness in one's research may affect one's reputation as a scientist, and might even result in the loss of a job, but it is a far cry from deliberate miscalculation or misinterpretation of data. Such a delineation—by intent—is well grounded in the law. An employee might be fired for sloppiness in cleaning a floor or even for damaging valuable equipment through inattention, but such behavior differs from willful or malicious destruction of property.

In publishing, errors in proofreading and typesetting affect a periodical's reputation for accuracy and detail, but those mistakes are not usually considered attempts to "defraud" subscribers. Likewise, an author's carelessness in calculation or statistical analysis, or mistakes in references, although offensive and inconvenient to colleagues, should affect the professional reputation but not trigger an institutional investigation—unless there is suspicion of knowing or willful negligence.[38]

Some authors, even those of considerable stature in a field, routinely submit manuscripts that contain "sloppy errors"; alert editors give those manuscripts special editorial attention and scrutiny before publication. In my experience, such errors follow predictable patterns—for example, wrong citations or inaccurate dates—and they signal authors who are rushed, are inattentive to detail, or suffer from poor eyesight, rather than authors who deliberately ignore their own errors. Although no one approves of this behavior, as Anthony Doob emphasizes, "a mistake does not make a fraud."[39] Evidence of deceitful intentions must always underpin the decision to launch a formal investigation, lest research that is controversial or that challenges orthodoxy too easily provokes a witch hunt.[40]

Drawing the line will never be easy. Standards change. They are also invariably affected by a field's internal politics and by shifts in power. Those authors whose experimental accidents appear in print should be embarrassed; those who publish their mistakes repeatedly should suffer from diminished reputations.[41] Neither type, however, should be thrust into the agony of a formal investigation of scientific misconduct without due cause.

CLASSIFICATION

Fraud, in the context of scientific communication, occurs when an author, editor, or referee makes a false representation to obtain some unfair advantage or to injure deliberately the rights or interests of another person or group.[42] In the publication of research, "fraud" encompasses false presentation of the authenticity of data, text, or authorship, as well as misrepresentation of a decision on publication; it includes deception intended to advance one's own career or position, to harm another's career or position, or to advance undeservedly another's career or position; and it includes theft of ideas. Table 1 classifies the types of actions most often

Table 1. Types of Unethical Conduct or Misrepresentation in Scientific and Technical Publishing

---

By authors:
- Describing data or artifacts that do not exist.
- Describing documents or objects that have been forged.
- Misrepresenting real data, or deliberately distorting evidence or data.
- Presenting another's ideas or text without attribution (plagiarism), including deliberate violation of copyright.
- Misrepresenting authorship by omitting an author.
- Misrepresenting authorship by including a noncontributing author.
- Misrepresenting publication status.

By referees:
- Misrepresenting facts or lying in a review.
- Unreasonably delaying review in order to achieve personal gain.
- Stealing ideas or text from a manuscript under review.

By editors or editorial advisors or staff:
- Forging or fabricating a referee's report.
- Lying to an author about the review process.
- Stealing ideas or text from a manuscript under review.

NOTE: This table draws on Daryl E. Chubin's earlier classification of misconduct within the entire research process.[43]

condemned in the context of scientific and technical publishing, apart from whether these acts violate law, custom, or moral principle.

Because each type of misconduct (e.g., fabrication, forgery, misrepresentation, or plagiarism) may be detected, investigated, and corrected differently for each type of person, I have chosen to classify the conduct according to roles in the publishing process. Many of these actions originate in a laboratory or similar unit, but they see the light of day as part of the process of communicating results to others. At the head of the list are those actions most well-documented and on which there is most consensus.

MISCONDUCT BY AUTHORS

Describing Imaginary Data, Artifacts, or Subjects

An author who knowingly describes data, artifacts, or research subjects that do not exist and have never existed violates an unambiguous standard of expected behavior. Such deception places the author at greatest risk of condemnation and punishment, yet is the most difficult to detect in the publication process because editors or reviewers do not normally ask to see original data or artifacts.

Ironically, to succeed in fabricating descriptions out of whole cloth, an author must know the research topic well enough to "create" (through imaginative description) objects, data, or phenomena that are plausible to experts (because they follow conventional theories, are statistically probable, or are apparently replicable).[44] Description of imitation data or artifacts must fit accepted paradigms of both the subject and the rhetoric or else there is the risk of attracting additional scrutiny.[45]

An author engaging in imaginative description must also be familiar with conventional publishing practices and know where the fraud could be vulnerable to discovery. Were the journal editor or referee to insist upon seeing data or other evidence cited in the text, this author could not produce it; but the risk of such a request is quite low. The deceiver (whether creating data within his imagination or creating a forgery for inspection) must therefore be "emphatically the child of his time . . . always in touch with the main current of ideas," and always sensitive to conventional practice in his field.[46] Sometimes authors have attempted to deceive by describing a mix of real and imaginary research; sometimes they have "improved" on an initial set of data or added false data to it.[47]

There is a long history for the description of imaginary data, ranging from eighteenth-century travelogues about nonexistent terrain or places the author never visited, to "anthropological" descriptions of imaginary native tribes or customs.[48] Sometimes these "expeditions" produced "data" that were relied on for years, or that gave credit where none was due. For example, suspicions have been voiced about Admiral Robert E. Peary's 1909 visit to the North Pole, a claim supported by his own notebooks and records.[49] Peary, in fact, attempted to refute the claim of his rival Frederick Cook, by accusing Cook of fraud. Determining the truth has always been difficult because Peary's records were "muddled" and there was a "lack of credible witnesses to the event"; but a new investigation has shown that some notebooks contain substantial errors and throw doubt on Peary's descriptions.[50]

Describing Forgeries or Faked Data

A second type of misconduct by authors includes the forging of documents or fabricating of objects (such as specimens, artifacts, or documents) or data books to support false claims. In such cases, the author describing the forgery can always provide them for inspection; they may even be sufficiently well-constructed to pass superficial scrutiny. There are two types:

new works that imitate or copy genuine objects but are made for fraudulent purposes, and genuine objects altered so that they appear to be something they are not (usually to appear older).[51] Instances of this type of behavior include the publication of an article that described research that was never done (e.g., in the Breuning case[52]) and a coauthor accused of manufacturing data after the fact, to support the conclusions of another published paper when its integrity was later questioned.[53] Fabrications of data need not always be massive and sweeping, either. Robert J. Gullis, a doctoral student in biochemistry at Birmingham University (1971–1974) who was also working at the Max Planck Institute in Munich, apparently manufactured results from only one phase of his research (at the final stage of his work), and there is no evidence that he did not do the remainder of the research. The fake was uncovered when colleagues could not replicate the work. Gullis eventually retracted eleven papers (seven of which were coauthored) and published a notice and apology in *Nature*.[54]

An author may have created the forgery or the fake personally or may have arranged for its manufacture. The ruse may also involve the entire object or just the manufacture of crucial missing parts. Many forgeries in history and archaeology have involved intricate pastiches, in which the forger intersperses fake and real parts, or else "shuffles" real parts from other sources to make a new whole.[55] These deceptions are obviously easiest when ambiguity is a normal part of the object's identity, as in archaeology or paleontology. Document forgers may create new pages to fill crucial gaps in an authentic text; archaeological fakers may alter or even deface an object to embellish its appeal or relevance. Paleontologists have been accused of deceptively "reassembling" skeletons, akin to taking one bone from "skeleton *a*" and another from "skeleton *b*" and then pretending they are from identical sources.

This type of deception tends to be most successful when the fake cannot be compared with an authenticated counterpart—for example, when the fraud involves either the "sole acceptable piece of evidence" (as in a faked photograph or geological sample) or some "necessarily rare or improbable" evidence (as in a forged historical document or a forged archaeological artifact).[56] Because of the skills required for successful manufacture, presentation, and purveying of a fake artifact, these deceptions have been called the "most grandiose" of frauds; many scholars cite the Piltdown skull as the exemplar, even though the identity of its forger remains unproven.[57]

The Piltdown skull was found in a gravel bed in Surrey, England, in 1911. For almost forty years, the skull was not only presumed to be authentic but also supported other anthropological and archaeological theories and was cited extensively by reputable scholars around the world. Yet, for all that time, the skull was never subjected to scientific tests, even though some experts publicly questioned its authenticity.[58]

The Piltdown forgery, like similar forgeries of unique objects, "rested on a datum which could not independently be provided by other scientists."[59] Its provenance seemed correct: the principal proof of its authenticity had been the accepted identification of the geological stratum in which it was found and the legitimacy of artifacts accompanying it, consisting of a plausible group of mammal bones, works of flints, and fossils. In 1953, when chemical and physical tests were finally applied to the skull, it was shown conclusively to be a forgery: the upper part of the skull was a genuine fossil, artificially aged to make it appear older; the simian lower jaw was a modern artifact.[60] The forgery's manufacture was relatively skilled for its time; the jaw, for example, had been "broken in just the two places that might have settled its relationship to the skull," and the piece was missing the chin regions and other crucial portions where it would have articulated with the cranium.[61]

Archaeology has been subject to fraud for centuries, perhaps because of the intimate links among the separate but entwined worlds of scholars, researchers, excavators, dealers, and collectors; but forgery or faking of artifacts and specimens may be found in many other fields as well (for example, Summerlin's "painted mouse" in biology).[62] In the 1970s, physicists and chemists would often argue that their fields were "practically immune" from this type of fraud; it was far more likely to happen where an observer's account of a rare phenomena is paramount or "where the true sciences merge into the humanistic studies."[63] Nevertheless, cases of faked experimental photos and samples have now come to light in physics and chemistry and as techniques of computer reprocessing of photographs improve, it raises the prospect of available means for subtle attempts at falsification in fields that rely on photographic evidence.[64]

When discussing this type of fraud in the context of publishing, it is important to distinguish between a forgery or fabrication by the author of an article, and a forgery manufactured by some other person but *presumed* by an author to be real. American archaeology recently confronted the implications of the latter type when questions were raised about the authenticity of the Holly Oak pendant, a whelk shell with a crude sketch of a mammoth or mastodon, which had been widely cited as "evidence of ancient human occupation" of the North American continent.[65] In 1988, modern technical analysis showed the shell to be "only a little over 1000 years old" and the engraving to be modern. When provenance or supporting theory is ambiguous, authenticity may remain in question for years. Scientific controversy continues, for example, over fossil remains discovered in a German quarry in the nineteenth century and later purchased by the British Museum, which some scientists believe to be those of *Archaeopteryx lithographica,* a bird that lived 150 million years ago and that would form a "missing link" between reptiles and birds. Equally strong opposition to the fossil's authenticity exists.[66]

If the authenticity of a data set or artifact has been formally challenged, authors who discuss that material in print have the responsibility to make the questionable status known, even if they still choose to reference it. Without such notification, an author would be perpetuating and magnifying the effects of a forgery. But what if there are only suspicions or rumors? And what if there is no way to verify authenticity? Until allegations are either confirmed or refuted, what should be the status of suspicious evidence? Art scholars argue that someone who unknowingly accepts a forgery and presents it as authentic is engaging in *false attribution* until the material is indisputably determined to be spurious (or genuine); thereafter, any failure to note its true status will be deliberate deception.[67] This analogy works well for science, too.

Sometimes, a forger deliberately seeks to have experts authenticate an artifact in order to increase its value. Perpetrators of this type of fraud may use the author of an article or book as the innocent disseminator of the fraud, as when an historian unwittingly describes a forged document or an archaeologist or geologist relies on specimens from a "salted" site.[68] Money is one motive but not the only one. Reportedly, a few manuscript forgers have sought to have their forgeries "touted as authentic" in order to bolster their egos or to seek revenge on experts who dismissed the forgers' previous work.[69]

A variation on both types of fraud may be found whenever an author has fabricated quotations from sources, either out of whole cloth or by altering real statements—for example, a sociologist, psychologist, or researcher in clinical medicine who fabricates or alters interviews or patient histories. In a recent ruling on the acceptability of fabricating quotations in a journalistic report, a federal appeals court made a statement about journalists which can be broadly applied to scientists who rely on interview data: ". . . truth is a journalist's stock in trade. To invoke the right to deliberately distort what someone else has said is to assert the right to lie in print."[70] Whatever the merits of that particular case, it raises issues applicable to all fields where evidence of direct quotation is used to bolster research conclusions. In scholarly work, quotation marks have been widely assumed to denote a direct and unalterable relationship to what was actually said.

Misrepresenting Authentic Data

Does the authenticity of artifacts or data insure against fraud? Certainly not, for authors can misrepresent or misinterpret authentic data and indulge in intentional research bias, that is, "the purposeful manipulation of an experimental design or results to confirm a hypothesis" or to illustrate an argument better.[71] This aspect of research practice attracts some of the most heated debate because it is the most ambiguous; there is little agreement among research fields about how far to stretch the definition of acceptable practice. Determining where unethical "adjustment" or "highlighting" begins and where allowable extrapolation ends usually depends on the specific research field. Moreover, from time to time, the demands of time and money influence standards for appropriateness, as can also the topic's urgency (e.g., the pressure to accelerate AIDS research, which has focused attention on redundant testing procedures). The exceedingly subtle distinctions between conventional practice and deceptive manipulation will usually be clear to practitioners but may be obscure to outsiders, further complicating disputes and institutional investigations.

For some research topics, "misrepresentation by omission" is not necessarily considered to be wrong because researchers adopt "a legitimate desire . . . to eliminate the irrelevant." "Awkward facts cry out to be ignored," they explain.[72] Sometimes omission is said to be the result of equipment or measurement "incapacity," or of researchers' "lack of insight" or "lack of imagination." The nineteenth-century British mathematician Charles Babbage referred to experimental manipulations as "trimming" (smoothing out numerical data) and "cooking" (selecting only materials that confirm one's hypothesis), his choice of terms emphasizing the deliberate nature of such representation. The continued popularity of Babbage's typology (it is frequently cited in the literature on scientific fraud) speaks to its appeal.[73]

In the social and behavioral sciences, an author might deceptively skew the selection of reports on interview subjects yet imply that all interviews are reported in full. In one case that was never formally resolved, critics have charged that a researcher deliberately withheld data that failed to bolster his case but instead emphasized what the critics call "unconfirmed experiments . . . cited . . . as 'personal communications'."[74] Even great historical figures in science are not immune to reexamination. Sigmund Freud, for example, now stands accused of data manipulation; historians of science report that Freud purposefully distorted case material and that he was, in essence, "an *a prioricist* who reached conclusions first and then doctored the evidence to fit."[75]

What happens if data are misrepresented intentionally, but only for a while? Is that unethical? Highly competitive research fields with potential lucrative commercial applications are often plagued by accusations that leaders in the field have deliberately misrepresented data in manuscripts in order to mislead rivals. One sensational allegation involved the substitution of the element Yb (ytterbium) for Y (yttrium) in a paper on superconductivity.[76] Although the symbol was changed before publication (but after peer review) and the author explained that it had been a typographical error, other physicists maintained that the "error" was a ruse designed to mislead competitors if the paper's conclusions had been leaked before publication and before patent applications had been filed. As the perceived value of discoveries (and resulting patents) continues to rise in certain research fields, such accusations will likely continue. No proof exists that any misrepresentation of this type has found its way into print, but many people are convinced the practice is growing.

Questions raised about the statistical methods chosen to interpret data may further muddy a controversy in ways not immediately comprehended by outsiders. Interpretive statistical practices vary widely among subfields and the task of review and detection of either mistakes or deliberate deception is daunting, given the volume of manuscripts received by most journals.[77] John C. Bailar, who advises medical journals on statistical review procedures, minces no words in condemning the "deliberate misuse of statistics" as an act of fraud, and he outlines three offenses: (1) when authors intentionally use an "inappropriate" statistical test "to exaggerate the significance of the results presented"; (2) when authors try several statistical tests but choose to report "only that which gives the best result"; and (3) when "hypotheses suggested by the data" are "reported with P-values, which have meaning only for hypotheses formulated independently of data."[78] For this category of misconduct, specific rules and policies must either reflect or be adjustable to the approaches of different fields and/or publications. However, all fields and all publications reject intentional violation of standards known to the author and willful violations of unambiguous standards.

Plagiarism and Deliberate
Violation of Copyright

Both the NSF and NIH now report that they investigate substantially more allegations involving plagiarism and stolen ideas than allegations involving falsified or altered data. Of the seventy-five allegations of all types of research misconduct received by the NSF Office of the Inspector General from 1 January 1989 to 31 March 1991, forty dealt with plagiarism or misappropriation of intellectual property.[79] Of the sixteen cases *resolved* by the NIH in a two-year period (from 1987 to mid-year 1989), seven involved plagiarism or stolen data, compared to nine involving faked or altered data; a later report from the Office of Scientific Integrity Review (a part of the Department of Health and Human Services which oversees NIH investigations) states that plagiarism occurred in five of fifteen investigations completed between May 1989 and December 1990.[80]

Plagiarism takes much the same form in all scientific fields and has been committed by researchers at all status levels or institutional settings, and in all types of communication. Its definition is simple—the deliberate presentation of another's texts or ideas as one's own or, as Alexander Lindey phrases it so well, "the false assumption of authorship, . . . taking the product of another person's mind and presenting it as one's own."[81] The reaction to plagiarism can be visceral; there is a curious element of intellectual violence even in the word's root [*plagiari(us)*, "kidnapper"; *plagi(um)*, "kidnapping"] because the abduction or appropriation of the words or ideas takes place without the owner's consent.[82] Even when "the plagiarist transfuses new blood into a decrepit body" and transforms dull ideas into shining prose, the unattributed appropriation that occurs without consent feels like assault to the original author, not rehabilitative therapy.[83]

The term plagiarism applies beyond the misappropriation of body text, to speeches and to the text of footnotes and citations.[84] When plagiarism involves wholesale republishing—without any citation or attribution—of previously published material, it may also violate copyright laws (see the discussion below). Like "pastiche forgery" of artifacts, plagiarism can include also a mixture of original text and rephrasings, perhaps with attribution, without quotation marks, and with the implication that the phrasings and interpretation are the plagiarist's own. And, finally, it can involve the presentation of another's innovations or inventions (or another's new approach to a standard experiment) as one's own.

Crucial to all definitions of plagiarism is some indication of an intent to deceive. Predictably, disputes lurk in the grey areas, somewhere between indisputable word-for-word, line-by-line copying of large amounts of text and the artful interweaving of original with plagiarized and unattributed text, for it is the deceptive aspect of plagiarism that has been condemned through the centuries, not the *use* of a predecessor's insights. Few scholars are so brilliant as to be truly original; we build our research on all that has gone before and rely on others' insights to point us toward new directions (or to serve as theories that we reject and disprove). Acceptable practice allows us to "use" another's words or ideas within our writings as long as we give appropriate credit—and most authors delight in those uses as measures of their influence. For writers whose work is interdisciplinary or analytical, in fact, innovation may come primarily through recombination of insights from different fields.

On its face, plagiarism is never considered tolerable in any field. Recent cases have elicited such strong public statements as: "As far as we and the public are concerned, plagiarism is a big sin, right up there with falsification of data," and, plagiarism is "at least as serious a misconduct in science as outright fabrication of data, if not more so"; and yet the scientific community has not treated allegations of plagiarism consistently.[85] When articles have been plagiarized "almost in toto" (as in the examples described in chapter 1), they have been universally condemned. When the plagiarism is something less, then attitudes of colleagues may be more lax. Scientists defending one accused plagiarist told reporters, for example, that they believed that "copying a portion of another's article" should be considered a "lesser offense than stealing the whole article."[86]

Some plagiarists have at first claimed that they were overly influenced by what they read; they played it all back verbatim, as if they had unconsciously recorded the words in their heads. Upon the discovery that his book included extensive verbatim quotations from other books, without quotation marks, an historian of science pled that he had "an extraordinary ability to remember material when I wanted to, but I have never before realized that I did it unconsciously"; he admitted to plagiarizing "as many as 55 consecutive words" in a total of eight pages of text but other historians claimed at the time that the borrowing was far more extensive.[87] The line between honest influence and dishonest appropriation must first be drawn within an author's own conscience and may depend on the extent of "interinfluence" on a subject. For example, two seminal works on literary plagiarism are Harry Paull's *Literary Ethics* (1928) and Alexander Lindey's *Plagiarism and Originality* (1952), both of which I read long before I read Thomas Mallon's *Stolen Words* (1989), which cites and acknowledges the earlier texts.[88] Paull and Lindey influence current thinking on plagiarism more by their *approach* than by their specific language (which now seems somewhat flowery) but those who want to discuss plagiarism do often quote them. And both Paull and Lindey, who strongly condemn intentional plagiarism, pepper their own texts with quotations and interpretations from the work of others, identified as someone else's words or thoughts but not always cited with a footnote, for example.

Each field of research, each institution, and each journal adopts slightly different criteria for attribution and citation, but new questions continually arise. What about such things as appropriating all or most of another writer's bibliography, or copying data tables? When are certain illustrations considered to be original contributions and when simply part of common knowledge? Should one credit descriptions of common experimental procedures or common phenomena? What about identical chronological descriptions of events? Don't we all tend to use more or less the same terms to summarize the plot of *Hamlet* or to define radioactive decay? When does something become a truism, become such common knowledge, that it need not be cited? Distinguishing between coincidental similarities in texts and the concurrence of thought and interpretation in a field, on the one hand, and the misappropriation of unusual language, original conclusions, or innovative descriptions, on the other, is essential for determining whether a suspicion warrants further investigation and the decision will inevitably depend on the individual circumstances of the author and the publication.

Interlingual plagiarism through verbatim, unacknowledged translation poses special difficulties for detection. In the example given in chapter 1, the plagiarist Bartholow included verbatim translations of Topinard's text instead of carefully stating which ideas or descriptions were his own and which were skillful English-language translations of Topinard's text. Translators may claim credit for the felicity of translation, of course, but not for "authorship" or originality of the ideas in that text.

To authors, these distinctions may be sources of continual confusion. A science publisher's *Guide for Authors* tells writers to "be scrupulous and careful in giving credit for material used from someone else's work." Regardless of the legality of the "borrowing" (that is, regardless of whether text adheres to the legal rule of "fair use" of copyrighted material), the *Guide* continues, an author should "acknowledge all material taken from another work and make clear which portions of [the] work comes from another source."[89] The principal nonlegal test is whether one gives credit appropriate to the form of publication, an approach that need not diminish the wit and originality of one's own synthesis, interpretation, use, or compilation, but that does stress appropriate acknowledgment. One may quote within a speech, for example, as long as it is clear who the original thinker is.

Publisher Eugene Garfield condemns an even subtler violation: to fail to acknowledge intellectual ancestors. He writes that "to cite someone is to acknowledge that person's impact on subsequent work. Citations are the currency by which we repay the intellectual debt we owe our predeces-

sors. . . . failing to cite sources deprives other researchers of the information contained in those sources, and may lead to duplication of effort."[90]
Garfield calls this failure "citation amnesia" (that is, "the use of ideas
without explicitly citing the source"); but such a gentle phrase obscures
the intensity of his rejection of this practice. Publishing allows scientists
to establish their claim to originality, he points out: "Once published, the
owner's claim to ownership is strengthened when the work is cited." When
authors fail to cite previous scholarship, Garfield believes they "default on
their loans."[91]

Another variation, one even more difficult to detect, occurs in computer
science when software programs or key routines are plagiarized. Two
scientists concerned about this problem observe that, in general, a pro-
grammer's level of technical sophistication will determine the approach:
"Novice, student plagiarism mainly utilizes certain stylistic and syntactic
changes," whereas expert programmers usually "introduce semantic
changes (e.g., changing the data structures used, changing an iterative
process to a recursive process, etc.)." Although the former type may be
noticed by a diligent reader, they observe, detecting the latter type requires
expert knowledge because the detective must be able to comprehend both
"the semantic and pragmatic structure of the suspect programs."[92]

Plagiarism of engineering (and architectural) drawings and inventions
presents yet other difficulties. In these fields, there is a wide grey area
between honest intellectual influence and blatant duplication. Similar ambiguities arise in disputes over appropriation of ideas from research funding
proposals.[93]

The concept of "originality"—in the sense of "newness" or novelty—
lies behind the publishing and scientific communities' strong condemnation of all plagiarism. In discussing criteria for rejection, Michael Gordon
points out that "an author should not be presenting research findings which
have already been published, or which are insignificantly different [from]
findings which have already found their way into print."[94] Marcia Angell
of the *New England Journal of Medicine*, suggests that repetitive publication of descriptions of the same research represents a level of "benign"
plagiarism, because the author rarely discloses the fact of the other publication voluntarily, and the second work often substantially duplicates the
first.

Useful analogies come from art history, where artists share ideas or
influence other artists but where the concept of originality still has meaning as "a creative, not as a reproductive or technical activity."[95] Alfred
Lessing spells out two "necessary conditions" for originality in art: first,
what he calls "particularity" ("a work of art may be said to be original in

the sense of being a particular object not identical with any other object") and second, "individuality" (the work "possesses a certain superficial individuality which serves to distinguish it from other works of art").[96] The argument that originality "rests on the fact that art has and must have a history" assumes that, without such history, "the possibilities for the creation of aesthetically pleasing works of art would soon be exhausted."[97] Artists "do not seek merely to produce works of beauty," but instead "to produce *original* works of beauty."[98] It is important in science, as in art, that "original" research extend beyond copying, that the researcher be doing more than simply repeating time-tested steps (as in a student laboratory exercise that reproduces a famous experiment). This characteristic also helps to distinguish repetitive publishing from plagiaristic theft of text or ideas.

Of all the literary-based crimes, plagiarism attracts some of the best analysis. Most writers on plagiary point toward certain common factors: notorious plagiarists are often accused of more than one instance of plagiarism, and there is usually not just one questionable passage per work but several.[99] It is this systematic aspect, rather than the inadvertent omission of a single set of quotation marks, that elevates plagiarism to the level of serious misconduct in research communication.

Misrepresenting Authorship

Authorship can be misrepresented in two ways, by including as coauthor someone who contributed to neither the research nor the writing, or by deliberately failing to list as coauthor someone who played a significant role in the research or writing. (Because chapter 5 concentrates on the debate over the responsibilities of authorship and coauthorship, I touch on these issues only briefly in this section.)

In some laboratories or research groups, it is common practice to add the name of the project supervisor or lab director to every paper published by a member of the lab, even when those supervisors have not been directly involved in conducting or interpreting the research. The practice is frequently defended as "customary." The International Committee of Medical Journal Editors (ICMJE) represents typical argument on the other side when it states unequivocally that authorship should be linked to participation sufficient to allow every author "to take public responsibility for the content."[100] That contribution, in the view of the committee, may come at any stage of a research project—design, data analysis, interpretation, drafting or revising the article, or granting final approval of the final text—but, in the words of the committee, "participation solely in the acquisition of funding or the collection of data does not justify authorship" and neither does "general supervision of the research group."[101] Either inappropriately adding or inappropriately omitting the name of a coauthor represents a type of unethical conduct, therefore, for it presents a false impression to the readers about who actually speaks in the article, and who vouches for its accuracy.

No one disputes that it is wrong to invent a coauthor altogether.[102] This practice is risky business, of course, and serves little purpose unless the real author is attempting to draw attention away from an impossibly large and largely fabricated output. In the recent well-publicized cases that have involved nonparticipating coauthors, the deception has often been linked to attempts to obscure some other misconduct.

Misrepresenting Publication Status

A type of misconduct that, on the surface, appears to be increasing and to be attributable to career or tenure pressure is the deceptive misrepresentation of a manuscript's status—that is, whether it has been accepted by a journal and, if so, whether it has been scheduled for publication in a particular issue (something that is normally done only when the publisher receives the final version). Misrepresentation of the publication status of a manuscript has been listed among "Other Deviant Practices" investigated by the Office of Scientific Integrity Review.

A letter to *Chemical & Engineering News* in 1990 complains that the "listing of 'publications' that are not quite published" is a "widespread" and "growing" problem.[103] The authors note that the terms "in press" or "accepted for publication" should apply only to articles that have been accepted and are in the production phase, not simply submitted to or under review—no matter how favorable the chances for acceptance may seem. Scientists have also been accused of repeatedly and deliberately citing technical publications incorrectly or, in their proposals to government agencies, of listing as "publications" papers that had actually been rejected for publication.[104]

These practices are risky because editors of academic journals are frequently asked to review government proposals, fellowship applications, and the like.[105] With computers, they are also avoidable; it is a simple matter to maintain accurate resumes that honestly reflect the current status of one's manuscripts or other works in progress.

MISCONDUCT BY REFEREES

Michael Gordon, in his primer on how to run a peer review system, mentions abuses by referees as one of the greatest concerns of authors and editors.[106] Despite such concerns, this topic has received relatively little attention from scientists or publishers, perhaps because it remains hidden behind a veil of ambiguity.

Lying in a Review

One type of wrongdoing involves the deliberate bias of a referee's report or deliberate misrepresentation of crucial facts in a manuscript review. A referee might lie for any number of reasons. Some may be personal, as in an effort to harm the career of the author or to advance the career of someone else. Other motives may be related to intellectual rivalries, as when a referee desires to discredit an explanation or interpretation other than the one he or she favors.

The referee who writes a report that inappropriately praises the manuscript of a friend, former student, or mentor without revealing the relationship deceives the editor, who trusts the report to be accurate and unbiased. Although no case has yet been publicly reported, some authors have alleged that competitors have deliberately written negative reviews of manuscripts in order to ensure delay of publication (if not also rejection).[107]

In another questionable practice, a referee may write a report on a manuscript but not read it at all. Sometimes senior professors will pass on a manuscript to a junior colleague or postdoctoral fellow with instructions to read the manuscript and write the review. If not otherwise prohibited by the journal, this practice may be acceptable as long as the editor is informed of what is happening and agrees to the change of reviewers. If the original referee signs the review and does not reveal the other person's participation to the editor, then that is clearly unacceptable behavior. Two important norms have been violated in such situations. By giving the unpublished manuscript to a third party without the editor's knowledge or permission, the original referee violates the confidentiality of the review process. Moreover, by signing a referee report he has not written, the senior scientist lies to the editor and misrepresents authorship of the review. Most editors welcome reviews from knowledgeable scientists, no matter what their age or formal academic status, but most require (or expect) advance consultation on any change in the expert evaluator.

Deliberately Delaying a Review

Although no published data exist to confirm or refute the assertion, many editors seem convinced that referees have delayed manuscript reviews with the purpose of either allowing the referee to publish first or slowing the publication of an old or new rival. Proving intentional delay is difficult if not impossible, but can be deterred by policies that automatically dismiss the recommendations of referees who return reports past agreed-upon deadlines.

Intentional delay may be related to conflicts of interest—for example, those stemming from financial arrangements with a competitor or the desire to publish a result or interpretation first. Although I do not deal directly in this book with the topic of conflict of interest, I must emphasize that sometimes conflicts are unavoidable during review. For research areas in which only a few academic departments train the majority of its scientists, personal relationships *will* subtly influence the refereeing.[108] Journals in many fields are developing and, in some cases, have already instituted policies to deal with these conflicts; open discussion of potential conflicts (including those involving money or commercial interests) can help to address problems before they become disruptive.

Stealing from a Manuscript under Review

Plagiarism by a referee of either ideas or text from a manuscript under review is universally condemned. In addition to factors in common with plagiarism in general, theft by a referee breaks the relationship of trust on which the peer review process depends. The extent of such violations of trust is difficult to measure. Alsabti apparently plagiarized from at least one manuscript that was sent to a reviewer but which Alsabti obtained surreptitiously; the proposal's original author discovered the theft only long after the fact (chapter 1). More recently, NIH sanctions were levied against C. David Bridges for allegedly plagiarizing ideas from an article sent to him for review (chapter 5). The abuse of confidentiality and trust inherent in plagiarisms from unpublished manuscripts make this misconduct particularly distasteful to scientists and publishers alike.

MISCONDUCT BY EDITORS, ADVISORY EDITORS, OR STAFF

Forgery, misrepresentation, and plagiarism may also be committed by an editor, editorial board member, or an employee of a journal or publisher. Because the conduct in question is rarely mentioned by explicit codes or regulations, problems are still usually handled informally and quietly.

Forging or Fabricating Reviews

For a chief editor or an associate, deputy, or consulting editor to forge or fabricate a referee's report violates expected standards of conduct in all fields. Such deception might be committed with the purpose of harming the career of one author, or unwarrantedly advancing the career of another; it may be done to advance a social or political cause. Whatever the purpose, the deception violates certain basic assumptions of appropriate behavior. No journal is obligated, other than by its own internal or institutional policies, to solicit expert review from independent referees; but the trust relationship between an author and a journal does require truthfulness about whatever review process is followed. In particular, the term "peer review" may not be used for review conducted by editorial staff alone unless they are proven experts on the topic and the author is informed that review has been exclusively internal, not external.

Lying about Review Criteria

Disappointment at a manuscript's rejection undoubtedly prompts some accusations of unfair editorial practices. Nevertheless, it is certainly unacceptable practice for an editor to reject or accept a manuscript for reasons unrelated to its technical merit, the space available in the journal, or similar legitimate editorial or administrative constraints. If a journal purports to evaluate manuscripts on merit, then its editors may not reject or accept a manuscript because of the author's race, gender, politics, friendship, or status, or because of a desire to impede the author's career. The integrity of the journal peer review system demands adherence to standards of research and scholarship, not obesiance to biased or hidden social or political factors. Here, too, there is little firm evidence that any journal or its editors have engaged in gross distortion of the review system, but rumors and gossip persist.

Among the most well-known allegations about the fabrication of referees' reports are those made about the psychologist Cyril Burt. During his sixteen years as coeditor and editor of the *British Journal of Psychology (Statistical section)*, Burt is alleged to have fabricated as many as half of the contributed reviews, notes, and letters published in the journal during that time.[109] Sometimes the forged reviews or letters responded to or commented on Burt's own work; by fabricating comments and publishing them under fictitious names, Burt apparently sought to have his work appear far more important than it was, for he then used the "reply" to these critiques to advance his own theories and interpretations.

Stealing from Unpublished Manuscripts

Plagiarism by an editor resembles, in its ramifications and variations, that by authors or referees, but it includes the element of abuse of trust and privileged position. In chapter 5, I describe the fictional events in Kingsley Amis's *Lucky Jim*, which closely approximate misconduct that has occasionally been alleged (but not proven on the record) about editors. The abuse of trust parallels that of reviewers who steal ideas from funding proposals.

DIVERGING PERSPECTIVES: "BAD SCIENCE"

The above list reflects the perspectives of those directly and officially engaged in the communications systems, within journal offices and organizations and in publishing companies. But one can also interpret the same behavior from distinctly different perspectives. Alexander Lindey (in analyzing plagiarism, but in a statement that applies more broadly) emphasizes that each part of a "clashing triad" of interests—art, law, and ethics—regards intellectual "borrowing" differently. For fraud committed during the communication process, four interests pertain—publishing, science, law, and ethics. It is symptomatic of the lack of consensus on research standards that, depending on the act in question, these interests can diverge rather than coincide. Ironically, the scientific standards have been the hardest to define.

No formal or written ethical "rules" govern the entire research process for science, engineering, and the social sciences; instead, written procedures tend to reflect each discipline's (or each institution's) special concerns about personal and professional integrity, about safety, or about public relations. Nevertheless, all fields share certain vague but identifiable concerns about, for example, the importance of accuracy or the wisdom of repeating critical experiments or calculations. At the practical level, there is greatest variation among the criteria for acceptable evidence, that is, the acceptable margins for statistical error or the amount and depth of evidence required. Anecdotal evidence or personal interviews, unverified by any second party, may be perfectly acceptable for one branch of psychology, for example, but unconvincing as primary evidence in another. Similar discrepancies arise in deciding when research is "complete," when there is sufficient data to prove a hypothesis or enough evidence to support an argument's conclusions. In one investigation, the accused scientists were criticized for, among other things, a lack of completeness in their research, leading a member of the investigating committee to remark sarcastically that his own experiences as an author and as a journal editor led him "to the inevitable conclusion that few authors would try and even fewer editors would accept an attempt to provide an account of a 'relevant' finding that every person in the world would consider 'complete'."[110] Some fields adopt conservative approaches to research practices; others reward ingenuity, imaginativeness, creativity, and intellectual risk-taking. Even the most daring scientist must adhere to his or her field's general world view, therefore, if he or she wishes to be accepted by the mainstream and avoid misunderstanding.

Despite the differences, some conduct is considered inappropriate in *most* disciplines. Certainly it is unacceptable to deceive human subjects about the risks of an experiment, although subjects may be deceived as part of an experiment if there is no risk to their welfare or well-being. Procedures in fieldwork or laboratory experiment may not endanger the health and safety of fellow workers or the public. No experimenter may invent or fabricate objects or data and then present them to colleagues as real. Likewise, all sciences condemn plagiarism "not only because it violates the scientific ethic of truthfulness" but also because, as one analyst emphasizes, "prestige, position and grant money are largely dependent on the originality and quantity of one's publications."[111]

Ambiguity and the individual requirements of different fields lead some scientists to define research fraud quite narrowly, as only "fabrication or deliberate misinterpretation of data" and (possibly) "deliberate theft of ideas." Others add "deliberate misrepresentation of authorship" to the list. The narrow definitions reflect the evolution of the debate over scientific fraud described in the previous chapter. In the 1970s and early 1980s, scientists tended to examine first what an accused scientist had done in scientific terms; colleagues and co-workers tended to look more at the importance of the research than at the individual's motives or the consequences of the misconduct. For example, they asked whether the accused had misinterpreted data, or just refashioned or improved on data? Had "a brilliant researcher" taken someone else's drab work and turned it into gold, thereby contributing greatly to science?[112] To some scientists, drawing an incorrect conclusion from real data represents a far worse crime than misrepresentation of data or authorship; to others, the "inadvertent plagiarism committed in a roundup of current knowledge in a field" is "less serious than in an article claiming to present original data, because a review article does little to enhance the writer's reputation."[113]

Approaching allegations by first investigating the scientific context retains control of the controversy in scientists' hands, but it can inhibit later formal investigation by initially obscuring potential violations of state or federal law. In a case reported in 1989, Stanford University uncovered apparent misdiagnosis of research subjects and incorrect classification of subjects, both errors that invalidated the study. The university was unable to assign blame for these problems, however, and unable to determine whether they were deliberate. Its initial investigation blamed the flawed study on "poor judgment and lack of communication among collaborators" and stated that it resulted from "a serious departure from acceptable scientific procedures."[114] Subsequent government investigation traced the tainted research to two specific members of the project team and found evidence of plagiarism.[115]

From the perspective of science, a false claim—even if it is made out of self-delusion (that is, out of wanting the idea to be true) or self-promotion (that is, wanting the project to appear to be successful)—violates a cardinal rule of cautiousness. Data, especially unusual, dramatic, or controversial data, should be checked and rechecked; an announcement of success or a new finding should not be made on the basis of one test. In practice, of course, a phenomenon may be so ephemeral that it is detectable only by a costly and/or time-consuming procedure, one neither easily nor quickly repeated.

Exaggeration and false claims, when they occur in print, quite frequently fall into the grey areas between deliberate fraud and innocent error. Ten or twenty years ago, a researcher with an otherwise impeccable reputation who had occasion to withdraw conclusions or to "retract" data may have appeared foolish or hasty, but colleagues would not have publicly accused him of unethical behavior. Today, the tolerance level has dropped, perhaps in response to political pressure. Although still by no means widely regarded as "fraud," the exaggeration or false claim attracts public condemnation now, even if it is not formally investigated. In one scientific about-face in 1988, a researcher at first claimed discovery and then, a few months later, admitted error, something countless other researchers have done. In this case, however, a few scientists publicly accused him of being overenthusiastic, overconfident, too hasty, and even of engaging in "showmanship and shoddy science."[116]

Attitudes to misconduct reflect scientists' overall opinions about the relationship of science to the community (local, national, or global) in which their research is conducted; these attitudes also reflect their sense of specialness. A scientist is a citizen of a nation as well as of the community of science; but many scientists sincerely regard themselves as different, as social mavericks, as political outsiders, as objective observers but not participants in society, or as nonconformists unconstrained by conventional mores or by laws.[117] Like other creative people, the moment a scientist "publishes his work he must reckon with the law—the libel law, the obscenity law, the copyright law," and so forth; yet scientists can sometimes seem to be ignorant of how those laws influence or affect science communication. Scientists, unlike accountants or attorneys or physicians, do not need licenses to hang out their shingles. Until the passage of federal regulations governing the use of human subjects and animals, few governmental regulations applied directly to how research in universities must be conducted.[118] The same lack of experience with the law inhibits understanding of the *public* nature of scientific publishing, that is, that scientific communications, although aimed at audiences of other experts, nevertheless disseminate scientists' ideas further (e.g., through the press, to society generally). Failure to recognize this connection also inhibits recognition of the public's interest in the integrity of that communication system.

VIOLATIONS OF THE LAW

Distinguishing between scientific and legal perspectives is crucial, because most of what is described as "scientific misconduct" does not violate any state or federal laws, and those laws or regulations that do apply are relatively new.[119] Even the use of the term "fraud" to refer to research fabrication, fakery, and forgery is problematic. In the United States, common law defines "fraud" as "an untrue statement of fact, known to be untrue (or made with reckless disregard for truth) and made with the intent to deceive, upon which another person [has] relied [and which resulted in their] subsequent injury or damage."[120] According to a strict legal interpretation of the terms, "injury or damage" to another person may not be "necessary to constitute scientific fraud,"[121] but the mere *possibility* of harm underpins political concern, especially the fear that medical treatments or bureaucratic decisions may be based on false research data.

Four types of misconduct common during the publishing process have clear-cut legal implications: (1) intentional publication of falsified or fabricated data; (2) deceptive authorship (e.g., as a violation of a contract between a prospective author and a publisher); (3) forgery of a signature (e.g., when a fictitious or nonparticipating coauthor's signature is forged on a publishing contract or submission letter); and (4) copyright violations, especially those related to plagiarism. None of these types of misconduct represent problems unique to scientific and scholarly publishing.

Fraud, Forgery, and Misrepresentation

The recent federal regulations on fabrication and misrepresentation of scientific data primarily concern statements made in proposals or reports to federal agencies. Presenting such data in a journal article may not necessarily be illegal—that will depend on the private contractual arrangement between author and publisher (e.g., if deception violates a publishing agreement) or on whether the article is used to obtain or justify the expenditure of federal funds, in which case misrepresentation has been held to violate federal law (see, for example, 45 *Code of Federal Regulations* 689) and becomes a public matter. These issues are still fresh and subject to differing interpretation, however, because relatively few cases have been prosecuted to date. In accounting, for example, falsification of data is clearly considered to be fraud.[122] If a scientist forges or falsifies data and then uses them to encourage private investment in a research project, then such behavior also has been considered criminal misrepresentation with intent to defraud. How these violations relate to the interests of private scientific publishers is less clear. Under Title 18 of the U.S. Code, Section 1001, submitting falsehoods (such as intentionally fabricated or misrepresented data) to the U.S. government in seeking or reporting on a research grant constitutes a crime punishable by fine and/or imprisonment.[123] At least one legal expert argues that plagiarism in a grant proposal or report might already be considered a crime under that statute.[124] The federal regulations implemented or proposed by NSF and NIH also include explicit prohibitions against plagiarism in proposals or reports.

In addition to a copyright violation, a plagiarism can be a contract violation. Contractual relationships with employers (or, for students, the promise upon admission to adhere to a school's codes of conduct—"honor codes") may be violated if plagiarism is proven. The standard provisions in the agreements that authors sign before publication of an article or a book also usually require them to warrant that a manuscript is an "original" work.

Literary Attitudes and the Law

Fear and loathing of plagiarism influences a wide range of publishing practices, but plagiarism is narrowly defined in the law. Literary scholar Walter Redfern refers to legal attempts to "pin down plagiarism" as "hilarious"—the word-thievery and intellectual purloining that have filled literature for centuries represent, in his view, a "legally unmanageable" problem.[125] In attempting to punish plagiarism, U.S. courts look first at who owns the intellectual property in question and/or who has a right to use or copy it (that is, who owns the "copyright").[126] The initial inquiry is thus not "who wrote it?" or "who is the author?"—these are aspects that are pertinent to determining the *moral rights* to original works; instead, the legal system seeks primarily to determine who possesses *economic rights* in the property. Next, the court determines whether the accused has "copied an essential or substantial portion of copyrighted or copyrightable material" without the permission of the owner (or outside the legally determined bounds of fair use) or whether the copying violates some contract.[127]

Originality, which is generally perceived as an artistic and literary virtue, is tied to the moral violation involved in plagiarism; originality may determine authorship but it is *ownership* that the law protects.[128] Some commentators, noting that plagiarism was not always regarded as a crime, trace the development of an economic-based definition to the growing acceptance of writing as a "trade" that produced a "commodity." As Thomas Mallon writes:

> [t]he history of copyright actually has more to do with piracy than plagiarism. Laws are far more useful in protecting authors against wholesale pirating of their books by publishers with no rights to them than they are in stopping the dead-of-night authorial theft of a passage here and a paragraph there.[129]

The development of copyright principles in eighteenth-century England in fact paralleled the notion of "individualized . . . literary creation" and signaled "the birth of the author-owned text."[130] Christopher Small points out that freedom of speech and freedom to print began to be "formulated together, both popularly and legally," and people "demanded them as though the first naturally included the second."[131] In the United States, copyright protection therefore encompasses economic (an author's and publisher's proprietary rights) as well as intellectual interests ("originality").

Both the concept and the law were formulated in a print-based era. English copyright statutes were constructed initially to protect publishers and booksellers against pirated reprints; those laws also attempted to limit monopolies on publication. Demonstrating its derivation in English law, Article I of the U.S. Constitution secures "for limited terms to authors and inventors the exclusive right to their respective writings and discoveries," and prohibits unlimited ownership by either authors or publishers. Definition of the specific terms and rights of ownership inherent in copyright have changed over time, but they have usually responded more to changing political environments or the development of new technologies than to any deliberate pressures to diminish authors' rights. Depending on whether you are an author or a publisher, current trends may seem either too restrictive (i.e., toward increased inhibitions on the freedom of users) or too lenient (i.e., failing to protect a publisher's investments).

Coincidentally, the most dramatic pressures for change stem from the advances of science and the needs of scientists. The Copyright Act of 1976, and subsequent amendments, have responded primarily to development of photocopying machines, as Congress has attempted to reconcile the changing rights and needs of authors, publishers, scholars, and educators with the possibilities of fast, inexpensive reproduction.[132] Especially in the area of scientific and technical journal publishing, the pressure to preserve freedom of use repeatedly clashes with pressure to protect commercial interests. Ever more periodicals, increased electronic indexing and dissemination, high-speed photocopiers and scanners, and an accelerating publishing pace in many "hot" research fields all coincide with normal economic pressures on publishing. Every photocopied article seems crucial to a researcher hot on the trail; every photocopy machine seems like a gun aimed at the publisher's heart. Vigorous defense of copyright has thus become the battleground for settling what may actually be disputes over economics. Another wave of change, discussed in the final chapter, will come with further improvements in electronic communication, networking, and text reproduction.

Copyright Law and Scientific Communication

From the author's perspective, the general spirit of U.S. law has always been well expressed in the famous opinion *Holmes v. Hurst* (1898): "The right . . . secured by the copyright act is not a right to the use of certain words, because they are the common property of the human race and are as little susceptible of private appropriation as air or sunlight; nor is it the right to ideas alone, since in the absence of means of communicating them they are of value to no one but the author. But the right is to that arrangement of words which the author has selected to express his ideas."[133] You cannot, therefore, copyright ideas, only their expression in print, and you cannot copyright news, facts, or general situations and plots. The interpretation of the precise boundaries of such prohibitions is continually changing in case law, and several good texts for non-lawyers, such as the current edition of William Strong's *The Copyright Book*, describe the implications of changes in the law.[134] In brief, however, to be copyrightable, material must pass a test of "originality": it must be original with the author and must be a "new" expression even if it is an old idea. Thus, the legal test examines the form of expression rather than the novelty of the subject matter and asks whether the work results from "independent laboring or copying."[135] The concepts of "authorial originality and authenticity" and the ability to determine same, Ian Haywood points out, underwrite the essence of copyright.[136] Because the law is "rooted in the idea that an author's work is the extension of his personality," efforts in the 1990s to disentangle authors' "economic rights" from their "moral rights" are forcing deeper consideration of the psychological component of authorship in order to determine apportionment of economic value.[137]

To understand a crucial aspect of disputes over plagiarism or falsification, one must also examine the essential rights to a journal article or issue from the publisher's perspective. Copyright secures a precisely defined economic right in a specific work, but the concept also supports a system in which publishers compete for customers.[138] Copyright law protects a publisher's right to operate freely in a "cultural marketplace" by enabling a firm to own or purchase exclusive rights to a work and thus secure a limited monopoly on its dissemination. For technical materials with small markets and for academic publishers operating on narrow profit margins, exclusivity can determine economic success.

Scientific and technical information, in the form of journals, books, reports, and data bases, is sold and purchased, marketed and ignored, just as is the text of other media, from pulp fiction and consumer guides to specialized telephone directories. In general, the law of copyright makes little distinction concerning content (with the exception of educational or "fair use" provisions in the United States) and makes no distinctions among publishers outside government.[139] (U.S. courts have made distinctions when content is deemed to be pornographic, seditious, or potentially damaging to national security, however.) Authors create properties (texts) in which publishers invest, but academic authors and scientists in the process of writing arcane journal articles usually do not think of them as commodities. The messy manuscript, the incomplete descriptions, and the unsorted bibliography loom as tasks to be completed. The law, on the other hand, makes no judgment of essential "worth" or "usefulness" (or even neatness or completeness); it does examine a passage in relation to the total work. In a scientific journal article, similarities to innovative conclusions or explanations may receive different consideration than similarities to standard descriptions of experimental techniques, equipment, or routines. The law protects all manuscripts equally, considering all as potential commodities for the marketplace, even the manuscript that is "bad science" or riddled with error. Thus the "owner" of a copyright (either the author, or the publisher if rights have been transferred) possesses economic rights in the work which plagiarism is considered to damage, either in whole or in part. The plagiarist becomes the subject of legal inquiry once he or she presents another's "commodity" as the result of his or her own inventiveness by publishing it (copying it) without authorization.

Applying such an approach, the law and the scientific community may find themselves in disagreement because the law "heeds neither good faith nor excellence of result," neither so-called good science nor bad science.[140] A court will be little concerned with whether the original source or the plagiarized work was a scientific review article or a novel, or with whether experts consider the manuscript's conclusions to be important, insignificant, or inaccurate.

From the perspective of those who use the information in journals, U.S. law codifies authors' responsibilities as well as their rights. In the 1976 revision, Congress formalized a judicial doctrine of "fair use," which limits the amount and frequency of use and examines its "purpose and character," the "nature" of the copyrighted work, and the "effect of the use upon the potential market for or value of the copyrighted work."[141] Users (including authors who quote other authors) who keep within the stated boundaries do not violate the copyright law, even if they violate moral standards by failing to provide proper attribution or citation for a quote, thus giving a false impression of originality. Works for hire, and "ghostwritten" speeches and articles, of course, routinely involve contractual arrangements whereby the real author relinquishes his or her right to public credit. In the laboratory, institutional procedures may also determine the extent to which credit is publicly apportioned.

The law applies to the failure to seek and obtain permission, a definable offense against ownership of intellectual property; it does not apply to violations of scholarly ethics or to deceptive presentation of material in order to steal professional credit. Only the owner of plagiarized material may initiate a lawsuit for copyright violation, and, in fact, this can determine whether unacknowledged appropriation of scholarly material becomes a matter adjudicated in court rather than in a university committee. The American Historical Association (AHA) points out in its "Statement on Plagiarism" (1986) that "civil action depends on the willingness [and ability] of the injured author or publisher to sue. Criminal cases arise only if the authorities decide to enforce . . . applicable statutes."[142] As a practical matter, the AHA notes, "plagiarism between scholars rarely gets into court. Publishers are eager to avoid adverse publicity, and an injured scholar is unlikely to seek material compensation for the misappropriation of what he or she gave gladly to the world."[143]

A newer type of legal dispute surrounds the use and deliberate misrepresentation of interview or archive data. Recent U.S. court rulings have questioned the legitimacy of paraphrasing certain types of copyrighted but unpublished text without the owner's explicit permission. These rulings appear to have increased authors' rights over publication, extending "the right of others to make use of an original document . . . through judicious paraphrasing or quoting of it," because until quite recently, judges had seldom objected "to limited use of authorized quotations, and almost never oppose[d] paraphrasing, providing that it [took] only the ideas and facts of an author and not his unique mode of 'expression'."[144] In *Salinger v. Random House*, however, a federal appeals court ruled (and the U.S. Supreme Court refused to reconsider) that a biographer's unauthorized paraphrasing of J. D. Salinger's correspondence represented a violation of copyright "even though the language of the paraphrasing often differed radically from that of the novelist" and the quotes were drawn from letters deposited in various archives and made available to scholars.[145] The limits on "fair use" quotation and paraphrasing are obviously undergoing new legal scrutiny, perhaps reflecting changing attitudes toward ownership of intellectual property and in ways that will surely affect intellectual property disputes in science.[146]

CHANGING DEFINITIONS

The inability to define what is expected (and its reverse: the inability to define what expectations have not been met, or what rules have been broken) influences every part of the controversy over scientific misconduct. It is most troubling, however, in the context of formal communication of science—the public face of research—where precision and accuracy are prized and where deceptions and mistakes may be preserved for all time in print.

Within the next decade, technological advances in electronic communication will help to redraw the boundaries of authorship and ownership and therefore to redefine accepted practice, publishing requirements, and eventually legal standards. Many observers predict that there is already a "winnowing away" of rights once securely established in the law. Whatever the direction of change, technological developments that affect the cost, speed, and boundaries of scientific communication will add new problems. If defining wrongdoing seems hard in a world of print, it will be doubly so in a world reliant on bits and chips. Policies and standards for every part of scientific publishing, from its managerial structure to its economics, will have to be reexamined because, as the next chapter discusses, they are all rooted in print-era attitudes and relationships. The sooner that participants confront the potential effects of technological change, both positive and negative, the shorter the period of definitional ambiguity may last.

3 Scientific Publishing
Organization and Economics

The discovery of knowledge without its communication leaves the
process of research incomplete.
—National Academy of Sciences[1]

Until quite recently, an allegation of unethical conduct lodged against the
author of a scientific journal article would have prompted considerable
behind-the-scenes discussion, perhaps involving the accused's department
chair, university dean, or research supervisor, who would not notify the
journal or publisher until the investigation was complete. An apology or
retraction would therefore appear months or even years after the original
publication. Until then, readers would have no reason to suspect the work's
validity. A conspiracy of silence surrounded research practices—no publicity, no attention, no immediate retraction, lest the public image of science
(or of the institution) be tarnished. Journal editors and publishers were
complicit in thus sanctioning discretion.

The publishing community's approval of such an approach to dealing
with even serious misconduct stems not from insufficient moral backbone
on the part of journals and their publishers (or even from fear of liability)
but from the special role that journals play within the research communities they serve. In how it is produced and printed and in how it is treated
by the law, a scientific journal is like a popular magazine, like any other
periodical. In its social, administrative, and economic aspects, however, a
journal in the sciences or social sciences may have a far more complex and
close relationship with its audience than do its mass-market counterparts.
The entanglement of interests works to the advantage of small, specialized
journals and their readers in that it provides political support and financial
backing for what may otherwise be marginal commercial investments. It
can wreak havoc when allegations of fraud or plagiarism surface.

This chapter introduces the business and internal organizational structures of publishing and some of the forces influencing how journals handle
allegations and investigations of ethical misconduct. Everything within the

publishing business pivots around one positive goal—to make public, to communicate, to spread words and ideas beyond the author's immediate circle. Understanding the subtle sociological effects of fraud requires understanding how the existence of deception undermines efforts to achieve this goal and introduces unanticipated costs, in time, money, and staff resources, that can affect the profitability of publishing overall and the future of marginal ventures.

HOW JOURNALS FIT INTO THE SCIENTIFIC COMMUNICATION SYSTEM

"Journal" is a generic term, applied in common usage to periodical publications of quite different physical characteristics, managerial structure, and frequency of issue. They may be weeklies, monthlies, or annuals, published at regular or irregular intervals. The sophistication of production varies widely, from computer-produced newsletters (and publications that exist only in electronic form) to several hundred typeset, glossy pages. Physical appearance of a publication is usually unrelated to importance within a field. Some journals bring considerable profit to their owners whereas others drain limited resources from nonprofit professional societies and the members' dues help to subsidize publication costs. The circulation of journals can range from a few hundred to tens of thousands, all of which affects economies of scale in production and, hence, subscription prices.

The common factor that allows all these publications to be considered "scientific (or technical) journals" is the characteristic least apparent to the casual observer and yet most critical for success or failure—the relationship they have with their audiences and the services they perform for them. A journal is a periodical that an identifiable intellectual community regards as a primary channel for communication of knowledge in its field *and* as one of the arbitrators of the authenticity or legitimacy of that knowledge. In providing these services, journals carry forward a tradition initiated by periodical publishers in the eighteenth century as they assembled accounts from disparate sources and presented them as a whole. Each issue of a modern journal resembles a snapshot of the "world" of knowledge in its field or topic—representative or evocative but neither comprehensive nor intended to be. In their ideal, journals do not just transmit information; they filter, evaluate, and unify it.

Journals also form significant parts of an integrated scientific communication system that includes books, technical reports, and oral presentations, and which provides the "products" for all types of research, industrial and academic, basic and applied. The thousands of publications produced annually by professional societies and by commercial and non-profit presses record in some permanent form the knowledge created through research. Once the documents (or other records) are created and manufactured, commercial or nonprofit organizations (e.g., research libraries and computerized abstracting services) then index, store, or disseminate them.

This vast system is uncoordinated but interconnected, a combination of public and private entities and interests. It crosses national borders and can render cultural differences among researchers insignificant, making it seem irrelevant whether the author works in Kentucky or Kiev. In the United States and many other countries, national governments support scientific communications to quite different degrees, some providing direct subsidies of writing and publishing; but the connective tissue worldwide, the "scientific communication system," is, by and large, a private-sector operation. This particular circumstance affects how journals and their publishers respond to allegations or investigations of research fraud and misconduct and, in consequence, raises some important questions about journals that are often taken for granted.

A BRIEF HISTORY

The concept of scientific journals dates formally to the seventeenth century, when a few scientists began to act as intermediaries through which their far-flung colleagues could communicate. In detailed technical descriptions, disseminated through epistolary networks, those who collected and studied natural history shared their work with others of similar interests. These informal exchanges of correspondence became institutionalized from the mid-seventeenth century, when Henry Oldenburg, secretary of the Royal Society, created *Philosophical Transactions* as a privately owned, profit-making venture. Historians and sociologists of science sometimes point to this publication as the forerunner of modern journals in that it contained descriptions of research which were transmitted, evaluated, and edited by colleagues. Harriet Zuckerman and Robert C. Merton, for example, identify the transformation during this time from "the mere *printing* of scientific work into its *publication*" as crucial, because there "slowly developed the practice of having the substance of manuscripts legitimated, principally before publication although sometimes after, through evaluation by institutionally assigned and ostensibly competent reviewers."[2] John Burnham places the birth of modern peer review much later, stating that peer review may have existed in isolation from the 1600s but that he has found little evidence of widespread adoption or institutionalization until the mid-twentieth century.[3]

The development of science journals coincided with the development of periodical publishing generally, exploiting a format well-suited to the role that professional communication plays among scientists.[4] Books took too long to produce and, for most authors, represented an unwieldy and expensive choice; pamphlets were too narrow for the breadth of work taking place. The periodical format, regularly issued and typeset, allowed fairly rapid and regular dissemination of research results to even relatively small audiences. Because this was also the time of development of professional societies and associations, journals helped to establish intellectual standards and to cultivate intellectual readerships.

The history of science often treats scientific journals as if they are the scientists' special artifacts, as inventions that, like the microscope, were devised to help them with their work. In fact, journals, whether in the sciences or the humanities, have been influenced most in their visual form, editorial structure, staff, organization, and economics by forces within society and by trends within the publishing industry generally. Their similarities transcend the particulars of content, for the same production techniques are applied to the production of journals on paleontology and particle physics, on algebra and radical sociology. In the law, in regard to copyright, libel, and taxes, a scientific journal resembles every other periodical offered for public sale. And escalating postal rates, interest rates, paper costs, and wages affect all publications, irrespective of content. As the "information explosion" began in the 1950s and 1960s, technological innovations in typesetting (including the introduction of computers and high-speed presses), lower commercial air shipment rates, and similar changes made it economically feasible to publish truly international periodicals, distributed to small, narrowly defined audiences around the world.[5] In the sciences, global operations now include not just conventional periodicals but also abstracting and indexing services, bibliometric services, and electronic library networks. In the 1990s, electronically produced and disseminated journals will further widen the opportunities for tailoring communications to specialized audiences as well as distributing mainstream publications more rapidly.

FEDERAL POLICY

In the United States, federal information policy shapes scientific and technical publishing, but it does so in an unconventional direction compared to how government policies shape science funding and training. Rather than increasing publishers' dependence on federal subsidy or strengthening regulatory controls, U.S. policies have generally encouraged scientific publishers to remain private-sector ventures. This independence appears to be a *de facto* result of a number of uncoordinated decisions. Various government reports, congressional studies, and contemporaneous observations from the 1940s and 1950s show that although U.S. policy makers frequently expressed concern about the health of the scientific communication systems that disseminate federally funded research results and, in a few instances, even acted to encourage growth, a lack of policy coordination at the national level and lack of political pressure to do otherwise allowed vast information systems—from print journals to data bases—to remain in private hands.

During the years following World War II, U.S. efforts to improve scientific communication focused principally on disseminating information from wartime research, with the intention of facilitating its commercial exploitation. This knowledge would then spur new industries and economic recovery. "What can be done, consistent with military security, and with the prior approval of the military authorities," Vannevar Bush queried in 1945, "to make known to the world as soon as possible the contributions which have been made during our war effort to scientific knowledge?"[6] One solution was to move information quickly into the private sector, so the wartime Office of Scientific Research and Development set up a Publication Board that had the authority to begin declassification and to expedite release and publication of scientific information.[7] The proposals then being circulated for establishing a "National Research Foundation" (which eventually became the National Science Foundation [NSF]) also included statements about the foundation's role to "promote dissemination of scientific and technical information," directed toward industrial and commercial application.

Once the NSF was created, its policies continued to treat journal publishing according to prewar assumptions, as a private professional activity of scientists but not necessarily as a matter for direct government involvement. In the early 1950s, when rising printing costs and other economic forces threatened the existence of several major journals and the NSF was asked for help, internal NSF opinion leaned against federal subsidy.[8] Although the agency gave modest financial support to a few operations, the NSF director and the National Science Board (NSB) members agreed that direct subsidies "should be given only temporarily," and then only to journals that seemed likely to survive. When the NSB reconsidered this policy in 1971, it reaffirmed that NSF should only provide "temporary assistance" to "primary" publication activities that were not undertaken by commercial enterprises.[9]

As a result of this attitude, the science communication "system" in the United States evolved as a conglomerate of private and public, profit and nonprofit ventures that overlap and interconnect. Subsequent attempts to coordinate even minimum federal policies on scientific communication have met with little success, in part because of the organizational fragmentation and because of the important role played by private-sector operations that do not speak with one voice. In the early 1950s, for example, a central information clearinghouse seemed to be essential for American scientists, and NSF officials initiated discussions to create it, but support was weak; fears of foreign espionage discouraged coordination of overlapping or duplicative efforts, and the clearinghouse idea failed.[10]

Other advisory bodies examined the federal role in coordinating scientific information activities in subsequent years. Each time, a coordinating effort was endorsed and sometimes even commenced; each time, it faltered, usually for political reasons. Even the fears of a damaging "information explosion" in science, fueled by the perception that researchers were failing to use efficiently the information being produced, did not spur reaction.[11] In 1962, Jerome B. Wiesner, the Science Adviser to President John Kennedy, called the "science information gap" more important than the "missile gap or the space lag," and he argued that because the profitability of scientific and technical publishing had increased, commercial market interests were unleashing a "flood of information" in science.[12] For many federal policy makers, the preferred response to such complaints was to consolidate federal operations and to develop greater electronic information handling capability (such as computer-searchable indexes). The National Bureau of Standards established its National Standards Reference Data System in 1965, and NSF also began collaborating with other federal information groups on these issues.[13] The approach to the problem of overall coordination and linkage remained inconsistent. The federal and nonfederal technical communication systems had by then "interpenetrated" and many analysts urged more sensitivity to relations between the two; nevertheless, U.S. policy essentially neglected the nonfederal system and either concentrated on the development of coordinating activities or failed to act. As a result, the overarching policy issues were also neglected. Issues such as how and by whom scientific information should be disseminated, what are the intellectual property rights of various participants in the system, who should control sensitive information, and what the standards of behavior should be, went unconsidered because there was no suitable framework or forum for discussion.

Federal attention to scientific journals has usually been enfolded into federal policy on all publishing; science attracts little special attention. If a commercial mass-market periodical folds, whatever its topic or importance, the federal government does not rush in with artificial subsidies. In that respect, the First Amendment fosters economic as well as ideological independence. When the NSF discouraged subsidies to scientific journals in the 1950s and, again, in the 1970s, it espoused the same basic policy. Minor support, through such things as allowing page charges to be paid from federal grants, serve as the only regular channels for government support and even these vary considerably among fields.

The lack of direct federal involvement promotes editorial and organizational freedom for scientific journals and their publishers, even though the public and private sector interests may be interdependent. Without an ongoing forum of policy making discussions at the national level, there has been little impetus to articulate fundamental rationales for free and open scientific communication, other than where such communication relates to national security. As the controversy over scientific fraud stimulates proposals for applying federal restrictions or regulations to scientific journals, it has brought a number of neglected policy issues to the fore. For example, does the size, extent, or importance of federal involvement in R&D efforts provide sufficient justifications for the government to treat a scientific periodical differently than it does other periodicals? How might proposals to monitor editorial or peer review of content affect journals' basic First Amendment rights? To frame responses to questions such as these, we must first identify the different roles that journals play in the life of science.

THE SOCIAL CONTEXT AND ROLES OF JOURNALS

A journal article rarely represents either the first or the only communication about a piece of research. It may be preceded by reports to employers or funding agencies, by informal and formal talks and seminars for colleagues, or by publication of "in-progress" descriptions. Nor do articles represent the last word: they form the raw material for books, review articles, and science journalism. This continuous chain of communication reflects the continuous nature of research, in which private and public phases commingle.

A single report on a research project goes through several phases (table 2), some of which may be simultaneous or overlapping. At each successive phase, those who read or evaluate the manuscript have increasingly fewer ties (professional, personal, or organizational) to the author, or to the author's organization, and the weaker the tie, the greater the objectivity of the evaluation is presumed to be. In the first phase (which may even take place while experiments or data collections are still going on), authors decide what the report will say and in what format. The next stage, the actual writing and revision, may involve one person or many, and can include participation by professional technical writers who have

Table 2. The Life of a Research Report

Phase		Involves	Tasks or Goals
A	Authorship (Creation)	One person or team	To decide what's included. To develop conclusions. To correct other work.
B	Writing	One person or team	To determine format and tone.
C	Internal Review	One person or many (may not have been involved in stages A or B)	To evaluate quality of writing and content. To determine publication's potential benefit or harm to organization.
D	External Review	One person or many (not involved in stages A, B, or C)	To evaluate quality of content and writing, apart from context.
E	Publication	Many people (not involved in prior stages)	To disseminate it to audiences not connected to stages A, B, or C. To document scientific knowledge for the future.

played no part whatsoever in data collection or interpretation. With computer networks and electronic mail systems, research teams frequently now subdivide the writing process or designate one member as the "writer." As drafts are produced, experts or supervisors within the organization check and/or possibly approve the manuscript. Only at the publication stage, however, will there be expert evaluators who are unconnected to the author and have not observed the research evolve. Eventually, publication will disseminate the work to the most detached audience of all, readers who "know" authors only through their writings.

Even though communication about a particular piece of research rarely occurs all at once, the peer-reviewed journal article is still widely regarded as the definitive publication, as the place where, to protect a competitive position or to establish priority, the author may reveal data or facts previously hidden. Journals represent the *principal* means of *formal* communication among scientists and social scientists through which research is made public and through which it is evaluated and authenticated by other experts, before and after publication.

Although discussion tends to focus on the evaluation and dissemination functions, journals play other roles as well. They facilitate communication among research disciplines and interest groups, either those who are formally organized in a professional society or those linked only by common intellectual interests. Journals unify the group's research outcomes by clarifying acceptable topics, approaches, techniques, and interpretations. They serve as public forums for the resolution of debate and dispute in research fields, and for evaluating scholarship. They provide an historical record of their group's organizational and intellectual development. They convey reward through prestige of publication. And they represent and confer power to those who participate in decision making.

The status of journals in each of these roles is both mythical and real, the hoped-for image of the former more often obscuring the practical limitations of the latter. Although most journals do all the things listed to some degree or another, it is the audience's *perception* of, and consequent response to, a journal (e.g., whether it is considered to be important) which determines its influence. Although quantitative indicators, such as the Institute for Scientific Information's "impact factor," can measure the relative effect of a publication's content through subsequent citation of the work, the audience's subjective reaction remains the main arbiter of status and reputation.

Communication and Unification

Another prevalent myth about journals declares that they represent the "voices" of particular disciplines. In truth, not even journals with the most inclusive democratic intentions represent other than an influential minority of readers and subscribers because so few members of a research field are actively involved in journal management or decision making. In an American Council of Learned Societies (ACLS) study of scholarly communication, Herbert Morton and his associates showed that over half of the academics surveyed in the humanities and social sciences had served as referees or editors, whereas slightly less than half had not.[14] Although 56 percent had refereed an article and another 25 percent had served as an editor, a considerable number had done neither. Comparable surveys of physicists or chemists would probably show higher proportions of involvement, but certainly not total participation for members of even these fields. Some economists point out that journals that depend primarily on library subscriptions in fact do not provide services to their subscribers at all but only to the small group of people who publish in them, "for whom publication brings successful certification, career advancement, and personal gratification."[15]

Publishers and journals actually "broker" communication between authors and readers, act as go-betweens to bring information to readers in accessible and convenient form; they disseminate work more efficiently, cheaply, and quickly than if every author attempted to do so individually. Were simple *communication* their only function, however, most journals would be in dire economic straits.[16]

They survive because social and intellectual *unification* is a central aspect of their significance. Journals help to "create an international sense of unity among scholars by connecting people . . . isolated geographically, politically, economically."[17] They also provide a sense of community, where authors share ideas and readers see what their peers are working on. Letters to the editor and book reviews facilitate the exchange of current opinion.[18]

This unifying role is reflected in how most scholars use journals. Half of the scholars in the ACLS survey reported that they "regularly" examined fewer than three journals, in addition to the four to five to which they subscribed.[19] The "average" respondent to that survey checked about a dozen journals, but monitored less than half of these in a way that allowed continuous reading of all content. Respondents admitted that it is "virtually impossible" to keep up, even at a minimum, with the literature directly in their own fields. "Monitoring" (e.g., to see who is publishing what) thus provided them with a reasonable and economical way of surveying their peers' work. Biologists, chemists, and physicists probably monitor even more journals; laboratories or research teams sometimes subdivide the tasks of sifting through print journals or electronic data bases for articles or abstracts of interest to the entire group.

Debate and Evaluation

Journals have historically served as forums for debates over technical issues and professional standards, either through their reports in news-editorial features or by arranging content to represent a spectrum of interpretation on unresolved controversies. Those journals serving large audiences usually allocate a proportion of pages in each issue to commentaries or letters to the editor.

Lack of consensus within a field exerts the most influence on content through the evaluation process. At one extreme, a journal issue may contain no content that has not been critiqued and approved by outside scientific or scholarly experts. At the other extreme, the content may receive no external review before publication; editorial boards or specialty editors perform all evaluation. A few journals accept every article submitted; others may commission the entire content of every issue. A mixed model is the most prevalent: some but not all manuscripts are accepted but all articles are subjected to peer evaluation and revision.

The evaluation process, however it is structured, represents the authority that legitimates the selection of content, especially for those readers who are not expert in the narrow topic at hand. The process awards "the presumed value of certification" along with the implication that articles subject to such vetting have added value and are theoretically "worth" more than articles not so certified.[20] Publication of an article is presumed to declare that a journal backs the manuscript with its reputation and has been willing to invest scarce resources (e.g., page allotments) as proof of its confidence.[21]

The extent to which a publication actually verifies technical details (e.g., by checking statistics or calculations) is directly related to budget and to authors' expectations (e.g., the *American Historical Review* concludes that its authors "depend on publishers for guidance" about citation practices[22]). At least one publisher believes this role to be related to social responsibility as well: editors have a "moral imperative" to insure accuracy and completeness, Eugene Garfield has written, and "are supposed to insure that relevant sources are cited in a paper submitted for publication."[23]

Archiving

The authentication process supports the journal's *archive* or storage function, to assure that each journal volume stands as an accurate record of its research field at time of publication. Implicit assumptions about the importance of this role in fact form the rationale for the practice of publishing errata, corrections, and retractions. A retraction notice attempts to say "never mind what was said," ignore this article, do not incorporate its conclusions or data into current wisdom. No article, of course, can ever be retracted, only repudiated. Yet the continuation of the practice perpetuates the notion that a journal's issues *en masse* stand as archives of truth and that "retracting" a spurious article sustains this archive's authenticity as a whole.

Critics have long argued that journals should never be expected to represent truth. Although many journals cross-check details such as footnotes, dates, or standard values, the volume of manuscripts submitted and the limited economic resources of most editorial offices prevent thorough checking. Even if they want to verify references, editorial staffs rarely have adequate resources to do so.[24] A survey of editors would probably show that most assign the ultimate responsibility for accuracy to authors, even though editorial offices continually check suspicious calculations, odd references, and so forth. The discrepancy between the myth (that a journal's editorial and review process authenticates every word, number, and assertion and that each volume of a journal represents a record of verified scholarship in that field) and the reality (that mistakes creep in at every stage of the process, from draft manuscript to bound issue and that no journal presumes to be perfect) poses few problems when the errors are minor, when they are caught, or when a correction notice is published in a subsequent issue. As long as every reader of the original article sees the correction notice, then the overall effect seems minuscule. No disputes arise.

When allegations of deception arise, however, these neat assumptions are turned upside-down. The precise question of who should be responsible for authenticity of the record, especially when coauthors are involved, becomes a point of dispute. The perception that editors should take responsibility for resolving disagreements over retraction has led to considerable bad feelings. Moreover, any attempt at change, such as independent initiation of a retraction notice, makes an assertion not just about the legitimacy of the particular article or about the culpability of the coauthors but also about the journal's reputation as a repository of truth.

Reward and Power

Appointment as an editor or as a member of an editorial board may come with standing in a field; but in a rapidly evolving subfield, a young, ambitious scientist, skilled at academic politics, may capture such a spot. At older, established journals, editorships tend to be awarded only after years of labor in the vineyard of professional service. Within medical publishing, some editorships are perceived as long-term jobs but there, too, credentials and standing in the field are indispensable assets. Correct or not, the perception that editorships equal professional rewards underpins the assumption that financial compensation plays little role. If the editor is a tenured faculty member, the university department may (but does not always) reduce teaching duties. Editorships, editorial board memberships, and refereeing experiences are routinely listed as part of one's service to the field in vitae and in applications for promotion and tenure, but they are often performed with no remuneration. Because editorships and editorial board appointments represent positions of high visibility, some people may use the positions to promote their own careers or intellectual approaches, of course. Others have quietly exploited the opportunity to advance the cause of groups traditionally disadvantaged in academe, such as women or people of color, or to encourage discussion of professional standards or ethics.

Whether wielded for self or society, the ability to decide what does (or does not) appear in print represents power that even the most arrogant editors do not take lightly. Try as one may to focus only on technical issues, one cannot ignore the human beings behind the manuscripts. Publication can help or hurt careers. Awareness of that potential impact weighs heavily on even the toughest editor when the author is deserving but the manuscript is not.

CONTROLLING CONTENT: OWNERSHIP AND MANAGEMENT

Administratively and organizationally, no set model governs relationships between journals and their scholarly or scientific audiences. Some journals are owned and operated by large scientific societies; others by commercial publishers; a few are independent and entrepreneurial, with loose affiliations to a professional group or area of study. Frederick Bowes III observes that a journal's higher purposes transcend the specific management or organizational aspects because the scholarly publishing system is both "the long-standing disciplined process by which scholars, displaced both in space and time, communicate" and "the major means by which the fruits of scholarship are disseminated outside the academic world to professionals and others hungry to apply the new knowledge."[25] The exact role one plays in this system, Bowes points out, shapes both perceptions and expectations—authors concerned (ideally) with "communication" and publishers with "dissemination." This orientation is generally true whether the goal is to make a profit (i.e., when more economic benefits flow to the publisher) or to break even (i.e., when benefits and costs are spread throughout the system).

For each type of publication, the owner not the audience hires or appoints the editor and editorial board. The owner also ultimately determines editorial direction and tone and either approves (or directly oversees) the process whereby content is selected. The owner determines policies on design, advertising, and circulation, decides management issues, provides the capital investment necessary for sustenance or growth, and absorbs any profit or loss. For commercial operations, the owner may be a single person, corporation, or group of stockholders; for nonprofit journals, the owner may be a professional society or a university. The British journal *Nature* is owned and operated by the commercial publisher Macmillan; its U.S. competitor, *Science*, is owned and operated by the American Association for the Advancement of Science, a nonprofit professional society. Journals can also be published by a commercial press (e.g., Pergamon) and owned by a professional society. Sponsors such as universities can play yet another role if they lend their names (and hence their reputations) to commercial journals in return for control over editor and content.

Some journals move through a progression of ownership arrangements across their lives. *Science, Technology, & Human Values (STHV)*, for example, was founded at (and owned by) Harvard University and initially published through the university's printing office; then it was co-owned by academic programs at Harvard and the Massachusetts Institute of Technology but published by The MIT Press (a nonprofit university press); later, the university sponsors contracted with a commercial publisher, John Wiley & Sons; and some years after that, the universities sold the journal to Wiley, which, within a matter of months, sold it to Sage Publications, its current publisher. Under terms of a separate contract with Sage, a professional society now appoints the journal's editor, with the publisher's approval, and the society's members receive a discount on subscriptions. *STHV* is now a commercial, for-profit venture, its success determined by market forces. Similar combinations of nonprofit "sponsorship" and commercial operation legal ownership may be found throughout scholarly publishing.

Learned societies and nonprofit professional organizations support hundreds of journals and regard publishing as part of the process of professionalization, dissemination of knowledge, and setting of standards. Learned societies, historian David Van Tassel writes, are "dedicated to the preservation, advancement and diffusion of knowledge," whereas professional organizations set "standards of professional performance" and "guidelines for professional training programs"; the journals of both types enhance communication among the members and encourage conformity to the group's standards.[26] Journals thus supplement oral presentations at scientific meetings by defining the range of acceptable research topics. For research areas that are subsets of larger disciplines, journals help to establish intellectual boundaries and define professional credentials.

Professional societies tend to control their journals more tightly than do commercial presses. The American Chemical Society (ACS), a professional organization founded in 1876, engages in publishing activities typical of the large scientific societies. ACS publishes dozens of journals, all of which are governed by the ACS charter, constitution, and bylaws. These society regulations include an elaborate process of appointment and review of editors, and seek to ensure that the journals reflect the society's goals and values. No one can direct an ACS editor to publish any particular content, but he or she must follow accepted guidelines for fairness and accuracy.[27] Each editor retains final responsibility for day-to-day judgments on content and editorial policy, but the ACS Board of Directors reserves the right to replace an editor for "questionable" or "unacceptable" judgment. The ACS regulations also state that "it is presumed" that all editors will use "peer review" and recommend that each manuscript be read by at least two reviewers knowledgeable in the specific topic or field. The editor is assigned the responsibility of assuring a "fair and equitable review" and administering the process.

Since 1979, the ACS Society Committee on Publications (SCOP) has regularly appointed ad hoc task forces to review each journal every few years. The task forces examine the "quality of papers published," the "appropriateness of their subject matter," the ratio of acceptance to rejection, the uniformity and fairness of the refereeing process, the extent of editorial control, and general production quality.[28] An editor knows that the task force is conducting a review and is the first recipient of its draft; editors may append comments to the final report to SCOP. Usually the review is tied to the editor's term of appointment, but a journal may be reviewed early if SCOP suspects that problems exist. The ACS system thus attempts to provide continuous performance monitoring rather than merely crisis intervention, a model followed by many other large organizations, such as the American Psychological Association (which brings its editors together for an annual meeting and publishes *Guidelines for Journal Editors* which are continually updated).

To most readers, the actual production processes appear uniform; most readers make little distinction between journals published by commercial, university, or scientific society publishers. Indeed, as Jack G. Goellner, director of Johns Hopkins University Press, observes: "University presses and commercial publishers . . . are almost identical—except, often, in scale and, always, in priorities of purpose."[29] A commercial publisher, he continues, seeks first "to make a profit and only secondarily to advance knowledge," whereas a university press seeks "to advance knowledge, and only secondarily to make a small profit, if possible, or at least to break even, or at the very least to operate within a budgeted deficit." For all presses, the decision to publish is an *investment* decision.

Each administrative arrangement affects editorial autonomy differently, however, and therefore will influence how editorial disputes are resolved. Edward J. Huth, editor of *Annals of Internal Medicine*, believes that scale is one determining factor—editors of well-known publications associated with large, affluent organizations cannot influence editorial policy overall as much as their compatriots at smaller journals, simply because the large associations and presses tend to separate a journal's business and editorial functions. If an individual journal represents only one among many in a wholesale operation, then, Huth argues, the editor tends to be less connected to the publication process.[30]

Another neglected aspect is the cross-national nature of journals publishing. Originally, the science publishing business was organized along national boundaries; then in the 1960s, more technical and scholarly publishers began setting up branches abroad.[31] United States firms such as McGraw-Hill and Wiley opened branches in Europe, and European publishers such as Elsevier (Netherlands), Springer-Verlag (West Germany), and Oxford University Press (United Kingdom) strengthened their North American operations. Other European publishers, like Reidel, North-Holland, and Pergamon, began to court scientific societies and international organizations to secure "long-term contracts for the worldwide publication" of monograph series and journals.[32] Mergers and acquisitions throughout the publishing industry in the 1980s further opened the international marketplace. But internationalization carries a hidden cost. Economies of scale may be good, but transnational corporate ownership increases the psychic distance between a journal editor and publisher. Differences in organizational roles can then be exacerbated by differences in cultural perspectives on, for example, ownership of intellectual property.

CONTROLLING THE BOTTOM LINE: LEGAL ISSUES AND FINANCES

The specific legal and financial arrangements for journals do not necessarily parallel editorial control. The concept of "ownership" of a periodical includes both real property (such as back copies, articles in type, subscription claims, and proprietary rights in the publication's name) and legal ownership of back copyrights.[33] In the early twentieth century, investors sometimes acquired failing periodicals just for their presses and typesetting equipment, but the list of current subscribers and the right to a periodical's good name still remain its most valuable assets.

Who invests in a journal determines who accepts the risks and liabilities.[34] Commercial publishers, for example, can "own" even those journals issued under professional society "sponsorship," because it is the commercial publisher who owns the lists, copyrights, and rights to the name and puts up the capital investment necessary for production. An owner accepts public responsibility for the content and has the legal right to cease publication. Contracts and permissions documents may relieve a sponsoring society from some liability, but as many editors discovered during the Stewart and Feder controversy, legal risks do accompany discussion of misconduct issues and can also affect the retraction process. Who sues or is sued may not be obvious to the casual reader of a journal, but it is of vital concern for editors and publishers.

The issue of who controls a journal has been pertinent in at least one fraud case. When a Boston University researcher, Marc J. Straus, was accused of fabricating data (or, that is, of encouraging the fabrication of it—for the charges varied as the case developed) in a project funded by the National Cancer Institute (NCI) at the university hospital, he maintained that he had taken "no part in the falsification and was framed by co-workers."[35] Nevertheless, because he had been the principal investigator, he signed an agreement with federal authorities in which he took responsibility. In 1982, after an agency investigation, NCI barred Straus from receiving funding from any part of its parent agency, the Department of Health and Human Services, for five years.

While he was under NCI investigation, Straus submitted a manuscript to *Cancer Treatment Reports*, a journal owned by NCI but supposed to maintain editorial independence from the institute. Privy to the status of the NCI investigation, the journal's editorial board (eight of thirteen of whom were NCI employees) required that Straus's raw data be independently verified, although such requirements were not imposed on other authors. After publication, it was learned that there had been a slipup in some part of this process and, again, Straus was required to have his data verified. In this case, the editorial board's ties to NCI and their knowledge of the investigation affected how they treated one author and his manuscript. Their extraordinary request was clearly intended to spare NCI from embarrassment, yet they did not prohibit Straus from attempting to publish his work. Now that tighter federal regulations and institutional procedures govern how accusations and investigations are handled, other federally sponsored journals may be placed in similar situations, where they feel forced to adopt more conservative positions toward authors accused or convicted of misconduct.

Legal contracts between publishers and sponsors determine who pays when copyright is to be defended, or what happens when there is malfeasance or negligence on the part of the journal or publisher, but similar arrangements in the future may include assignment of costs for investigation of related accusations of research fraud or forgery because the most important factor in journals publishing is the bottom line in the account books. Economics affects policy as well as operations in that "the owner who pays for publication of the journal has the ultimate legal authority in setting journal policy."[36] As one publishing executive noted, "It takes money to establish and operate the kind of infrastructure that enables scholarly communication to take place, and it takes money to select, synthesize, package, and disseminate knowledge."[37] When a case of ethical misconduct arises, one set of factors influences who takes moral responsibility for an investigation, another determines who must accept financial responsibility for any associated legal costs.

At the editorial level, the operations for a journal with a circulation under a thousand can seem small and insignificant compared to *Time* or *Newsweek* but, collectively, journals are big business. The typical serials budget of a large research library runs in the millions of dollars. Louisiana State University, which has 26,000 students, one-quarter of whom study some type of science, spends about $1.6 million every year for serials.[38] Specialized engineering and science libraries, such as those at MIT or California Institute of Technology, may subscribe to over 20,000 different serials (including abstract and index series); Georgia Institute of Technology's library system received over 28,805 serials in the 1986–1987 academic year. In 1988, the libraries who comprise the Association of Research Libraries spent about $222 million on periodicals of all types.

Specialized journals are rarely supported only by subscriptions; instead, revenue comes from subsidies, grants, page charges, advertising, and (if owned by a professional society) dues payments from members. For society-owned journals, the members represent a stable and predictable circulation base that can be augmented by nonmembers who pay higher rates. Some societies try to keep marginal journals afloat by setting membership dues on sliding scales. In 1981, for example, the *Journal of Biological Chemistry*, owned by the American Society of Biological Chemists, published 12,000 pages annually. The journal had 7,000 subscribers, 5,000 of whom were not members of the Society and who paid a special rate ($285/subscription); each of the 1,200 members paid $100; 650 student members paid a nominal fee. Authors paid page charges of $35/page but the *Journal* did not pursue those who failed to pay; nonmember subscribers thus paid for approximately 75 percent of the cost of the journal. This is a practice followed by many periodicals—to charge institutional subscribers, libraries, and nonmembers a higher rate than individual members.[39] Nevertheless, when a university press or other nonprofit publisher decides to cease publication of a journal, the decision may be based on very slender cost margins. The "break-even" circulation levels for scholarly and scientific journals fall in the range of 500 to 1000, depending on the cost of production.[40] With such small numbers, it is easy for controversy to swell the circulation or to kill off a struggling publication.

Advertising may also be affected by controversy. Support derived from advertising can account for as much as half of a journal's revenue, and many nonprofit scientific societies rely heavily on advertising revenues from their journals. For example, *Science* earned $12 million in 1987 and over $13 million in 1989 from its advertising alone; that figure represented the largest single source of revenue (between 30–40% of the total) for its owner, the American Association for the Advancement of Science (AAAS).[41] Similar situations obtain at other large "umbrella" scientific organizations, such as the Institute of Electrical and Electronics Engineers (IEEE), the American Chemical Society (ACS), and the American Institute of Physics (AIP), which publish dozens of revenue-producing journals. Currently, their advertising revenue is taxed in the United States at a rate favorable to nonprofit organizations, allowing them to offset advertising revenue against editorial costs.[42]

Through page charges or manuscript "handling fees," authors also help to support editorial operations; for a single article, fees can exceed $1,000.[43] Revenue structures are somewhat different for European journals, especially for those produced in Great Britain, because they usually depend far more on subscription income, drawing only modest income from advertising, offprints, and back issue sales.[44]

Whatever the proportions, a significant loss in income from any one of these sources can jeopardize the survival of a small journal. Because researchers regularly read or monitor dozens of journals—sometimes from several other disciplines—no single discipline or specialty really supports the entire formal publications system that its members use. In fact, most of the immediate financial risks are borne by the commercial or university (nonprofit) press publishers of journals, and by sponsors' long-term underwriting.

The health of the journals market was tested in the late 1970s and early 1980s, when periodical prices rose dramatically in many scientific fields. The increases precipitated an "acquisition crisis" in research libraries, as they felt forced "to reduce drastically or to eliminate entirely the purchase of books in order to maintain journal or other serial subscriptions."[45] Some libraries reduced the number of periodicals overall. The combined libraries of the Smithsonian Institution, for example, which cut across every academic discipline in the sciences, arts, and humanities, experienced a 46 percent increase in serial costs in the late 1980s, and so the libraries canceled hundreds of journal subscriptions. Elsewhere, cuts went deeper. One university librarian has described a meeting at which 551 subscriptions were canceled, on top of 500 that had been canceled in the previous twelve-month period.[46]

As prices escalated through the 1980s, librarians began to express considerable antagonism toward the journal publishers; they began to voice consumer-like complaints and point to what they perceived as declining quality in the journals. Controversy over research fraud played right into that argument. Although the average cost per serial (for those purchased by major research libraries) was around $100 (estimates varied from $75–115) in 1989, subscriptions for a few highly visible journals cost in the thousands of dollars. Dramatic increases in institutional subscriptions became annual events in "hot" research fields. In computer science, one prominent journal rose from $425 in 1983, to $1,425 in 1989; in bio-medicine, another went from $160 in 1984 to $540 in 1989.[47] Publishers explained that the increased subscription prices reflected increased costs for such things as postage, paper, and labor.[48] The rapid proliferation of specialized journals, the movement of their ownership from nonprofit societies to commercial publishers, and differentials in international markets also influenced journal economics.[49] There is little consensus on the root cause, only a perception that prices have risen too fast, too far.

Others see the situation quite differently. William Broad believes that "the root of the publication problem" is not paper or postage costs but the fact that the journal publishing system is "carefully protected from market constraints."[50] In some areas of biomedical research, "[a]lmost every article . . . eventually finds its way into print," he asserts, because the federal government acts in ways that regard all biomedical literature as "suitable for publication . . . suitable for government support."[51] As evidence of a protected market, Broad cites the use of federal grants to reimburse page charges or reprints and university overhead payments that support libraries and, hence, institutional subscriptions.[52]

Whatever the cause, a secondary effect of dramatic price increases has been heightened agitation for quality control.[53] Fraud of any type is perceived as potentially destructive of quality. At the 1989 annual meeting of the Society for Scholarly Publishing, Hendrik Edelman of Rutgers University voiced opinions shared by many librarians when he urged publishers "to examine the mission, objectives, and performance" of each journal; other speakers in the same session demanded that, given the high costs of subscriptions, publishers should guarantee "fraud-free" products.[54] As subscription prices continue to rise, affecting market shares and increasing the perceived value of article publication, this debate will continue and strengthen.

PUBLISHING TRADITIONS

Curiously, one of the least discussed aspects of journals publishing is their responsibility to the communities they serve, a silence related to ambiguity about what that responsibility should be. Before she can investigate an allegation, the editor who suspects deception by an author, for example, must first decide to whom she is accountable, legally as well as morally, and whose rules should be applied to the conduct in question. Where does duty lie? To the accused and to his or her professional reputation? To the accused's institution? To federal sponsors or legislators? To subscribers who have paid for what they trusted were "authenticated" results? To the publisher who wishes to uphold the standards of good scholarship and to protect copyrights, and who believes deception should be punished? To the journal's primary constituency, embarrassed by the incident but convinced that it does not represent widespread deviance? Ideally, these interests should converge. When they do not, it may be because the norms for publishing differ from those traditionally assumed for research.[55]

In publishing, powerful standards of *trust* and *loyalty* affect the relations between author and publisher. To publishing executives "the mere idea of fraud [is] a nightmare dismissed to the recesses of the mind"; the industry is "still tinged with gentlemanly virtues like trust and friendship and loyalty."[56] Moreover, publishers tend to regard that loyalty as reciprocal. If a writer loyally stays with one publisher, then the publisher is loyal in return.[57]

Honesty and the expectation of honesty play crucial roles in this loyalty. Editors expect authors and referees to be honest with them; likewise, authors trust editors and referees. It's a pragmatic assumption, however. The system simply cannot function without mutual trust among its participants, because most interactions occur by mail, telephone, or fax, between strangers who may never meet face to face. John Dowling, a scientist and an editor, observes that editors need reviewers "who are in the best position to assess" an article but those people are often also in the "best position to take advantage of such information," so trust and honesty are essential.[58]

In the ideal situation, *objectivity* and *fairness*, not personal biases, govern editorial decision making, and decisions are unaffected by personalities, grudges, or gossip. "Outstanding reviews should not be biased by personal knowledge of the authors," one editor writes; "equal treatment for all authors is the goal to which we should aspire."[59] In setting out an "ethics of knowledge" for university scholars, Clark Kerr advocates a combination of tolerance, reliance on "intellectual merit" rather than personal factors, and openness about any personal evaluations that may take place, even as he acknowledges how difficult such goals are to realize.[60]

Robert Merton, in his sociological studies of scientists, proposes that they, in the ideal, evaluate new knowledge critically and objectively, holding their judgment in suspension. This expectation extends to the editorial process in science, which must initially view manuscripts with *skepticism*, making some provisional judgment about suitability.[61] Thorough skepticism obviously is impractical.[62] No editor wants to have to approach every article, every piece of data, as if it were untrustworthy; some limits to skepticism must exist.[63] Discovery of fraud inevitably makes everyone wonder whether they have been skeptical enough.

Ideally, an editor espouses *intellectual altruism*. Applause for an article goes to the author; by tradition, the editorial process remains invisible.[64] The actions of other advisors and referees are similarly expected to be guided by beneficence toward the authors or concern for the journal's best interests, to ensure its *success* and *survival*, not endanger its reputation, profitability, or existence. Editorial decisions are frequently made with an eye to financial stability, for example, or made to preserve a journal's continuity with its intellectual foundations. Whether the goal is profit or evangelistic expansion of a discipline, journals still must increase or maintain their circulations, or die. Promotion and marketing, either for money, subscribers, or intellectual colonialism, may therefore be an acceptable motive for editorial action. As an example, publisher William Sisler mentions such economic tradeoffs as choosing a manuscript that is "hot" or "trendy" even if one finds it "personally or politically distasteful."[65] For scientific books and monographs, the tradeoffs are sharper and the risk of mistakes even greater; for journals, the negative effect of any one article is spread over an entire issue. Thus, commercial or mercantile values inevitably influence (or are a factor in) editorial decision making in scientific publishing even if the publisher is nonprofit.

Three other norms frequently conflict: *openness, private ownership of intellectual property*, and *discretion*. As Diana Crane observes, "The scientist does not own his findings"; personal secrecy is not generally condoned by the scientific community "beyond the time required to present [data] in defensible form."[66] Publishers support an open marketplace of ideas from which to select content but also strongly supports private ownership of text and expression of ideas through the copyright laws. An editor may be opposed to secrecy in the publication of research results and still support private ownership of copyrighted material and protection of individual rights. Similarly, a standard of discretion affects all aspects of editorial operations. In an endeavor to avoid harm, journals expect their staffs to be discreet and to maintain the confidentiality of editorial recommendations and evaluations; moreover, they extend such expectations to referees and board members.

Originality is also much prized in publishing; reprinting draws less reward than printing "new work."[67] In this respect, journals adopt the norms of their discipline—seeking not just "good" reports of research, but reports that are distinguishably different from those published elsewhere.

Finally, journal publishing is ultimately a *collective* activity. Although a manuscript may begin in solitude, one author alone with her thoughts, the act of publishing is cooperative. Publishing gives expression to an author's desire to share information with others, and turning ideas into the printed word requires enormous collective effort by dozens of people in the editorial office, publisher's headquarters, typesetters, printers, binders, and so on.

Editors sometimes assume that all who come in contact with the publishing system not only should but also do adopt these standards, and that they automatically govern the actions of referees, members of editorial boards, auxiliary editors, and authors. The norms, however, are rarely expressed publicly. As a result, the standards and expectations for conduct, as well as how they relate to the legal and economic factors in publishing, receive inadequate scrutiny and critique. No one really notices these norms until fraud exposes their violation.

4　Authorship

Forgeries . . . have a habit of exposing dubious practices.
　　　　　　　　　　　　　—Ian Haywood[1]

Skim milk masquerades as cream . . . Black sheep dwell in every fold
. . . Bulls are but inflated frogs.
　　　　　　　　　　　　　—H.M.S. Pinafore [2]

In the 1980s, many of the investigations of research falsification cases revealed a curious and disturbing practice. The accused scientists had systematically involved unsuspecting colleagues and co-workers in the deception by listing them as coauthors on articles describing the faked research. Moreover, when the articles were discovered to be based on fabricated data, some of these "honorary coauthors" acknowledged that they knew their names had been used but disavowed responsibility for the content. In case after case, senior scientists as well as hapless postdoctoral fellows claimed ignorance of what had happened. They had not suspected fakery. They had been deceived. They, too, were victims and should not be held accountable for someone else's crime.

The practice of awarding honorary coauthorship was neither new nor surprising. No journal encouraged it, but most appeared reluctant or unable to prevent such listing. The controversy over fraud and fabrication just drew attention to how extensive the practice had become. Once it had been discussed in university investigations and castigated in congressional hearings, everyone seemed to agree that the concept of "authorship" in scientific and technical publishing needed careful reexamination.

That reassessment has not been easy. Political debate has dragged private relationships between authors and publishers into the open, testing the trust that glues the system together. It has forced journal editors, university officials, and policy makers to define "authorship" in order to determine where violations have occurred. In the process, it has been discovered that finding one definition of authorship acceptable to all institutions and all disciplines is like trying to grab the Cheshire cat—not just "intellectually difficult," but impossible.

OFFICE PROCEDURES

An author and a publisher (or an editor acting as the publisher's official representative) enter into two parallel relationships when a manuscript is submitted, one legal and the other social. Publishing agreements or contracts formalize the legal relationship: an author attests to being the "true author" of the material submitted, legally able to give permission to publish. A "social contract," a form of unspoken understanding, governs the other relationship and influences how journals react to and handle suspicions of misconduct by authors.

Both the legal and social arrangements rely on assumptions that are rarely articulated and even more rarely questioned. As a practical matter, publishers and editors tend to assume that each author is who he/she says he/she is, that all authors are listed, that all authors listed were authors of the work, and that the authors listed actually did the work described, but no journal or publisher has the resources to verify any of these assumptions for every article submitted. A journal must also trust that the paper's conclusions fit the actual data and that the article describes research that was actually done. Its expert advisors can only evaluate the technical accuracy of the data, statistics, or conclusions included in a manuscript, not the integrity of the research it describes.

When manuscripts are received, most journals follow a proscribed procedure: logging in and acknowledging receipt and then conducting initial review to eliminate incomplete or *prima facie* unsuitable manuscripts. The initial screening, sometimes conducted by professional staff, determines the manuscript's major topic for assignment to content (or associate) editors or to peer reviewers, but it rarely assesses other aspects.[3] Even though most interactions with authors happen by mail, fax, or computer modem, each submission is treated as if an author had personally handed it to the editor, saying, "This is my work and to the best of my knowledge it is correct." An editor has few means to verify that statement.

"AUTHORSHIP" BECOMES AN ISSUE

The discovery, during Harvard University's investigation of the John Darsee case, that many of Darsee's coauthors were less familiar with his research than might have been presumed, shocked both scientists and nonscientists alike. When episodes of similar negligence followed, editors and research administrators joined in calling for reconsideration of who should properly be listed as an "author" or "coauthor" of a scientific paper. This attention coincided with changes in the research climate in many fields (such as high-energy physics and molecular biology, where big teams and large laboratories were supervised by scientific "stars" uninvolved in routine activities) and with concern that aggressive competition or lax supervision might be contributing to failures of responsibility at both the laboratory and the individual levels. All too often, people just seemed to be taking credit for work they had not done themselves. Graduate students and postdoctoral fellows were expected, in some laboratories, to be like sharecroppers, to feel beholden to the senior scientist who "graciously" allowed them to join his laboratory or to be employed on his government-funded project.

In that context, in 1985, the president of Stanford University, Donald Kennedy, chose "academic authorship" as the topic of an open letter to his university's academic council and faculty. The essay received national attention in the scientific press and was reprinted and distributed widely the following spring because it touched on two emerging concerns: first, the perception that the university research climate itself had somehow made it "difficult to determine responsibility of authorship" and, second, the awareness that research groups often lacked explicit guidelines for allocating credit and responsibility.[4] Plagiarism (presentation of the "methods, data, or conclusions" of another as one's own) and copyright infringement were violations with which universities were accustomed to dealing, Kennedy observed; but the failure to acknowledge a coauthor or the inclusion of a nonparticipating coauthor represented dilemmas not covered in existing regulations. In the past, disputes had been adjudicated privately; now universities were being asked to adopt a more active and public posture.

Different disciplines often accept quite different coauthorship practices. In some fields, graduate students work independently; in others, they work under close tutelage. A determination of professional misconduct should take such variation into account, Kennedy argued, but all fields adopt some criteria for authorship and make some common assumptions. He proposed a list of principles applicable to all fields. First (and most important in a university setting) is the supervisor's responsibility to the student, that is, to train and to educate, not to exploit. Second, the "seamlessness that often characterizes collaborative research" may increase the difficulty of retrospectively dividing credit and responsibility; therefore, discussions about responsibility, credit, and authorship should take place before research begins. Third, in other types of writing, such as review articles, book chapters, or collective works, coauthors should clearly distinguish their respective contributions in appropriate citations. And fourth, a "tight coupling between authorship and responsibility" tends to bubble up when research has been shown to be fraudulent. In such cases, attempting to translate "minimal participation" into "minimal responsibility or blame" will not work if one has claimed maximum credit as a coauthor. To Kennedy, "each coauthor of a work is accountable for its authenticity and quality." "Shared credit should entail shared responsibility," he emphasized.

DETERMINING AUTHORSHIP: PRACTICAL MATTERS

In science, credentials provide a most rudimentary check on authorship. Most publishers use them as guidelines, tending to assume that authors tell the truth. Few editorial offices cross-check affiliations with *Who's Who* or similar guides, or with university personnel offices, but even such diligence would not catch every falsehood.

The practical obstacles to verifying credentials were demonstrated clearly in the E. A. K. Alsabti case (chapter 1), for Alsabti had had only tenuous, temporary connections to many of the institutions he had listed as affiliations.[5] One of the scientists whose proposal Alsabti apparently plagiarized has since suggested that editors should verify the credentials of all authors of review articles, especially if authors "have never published original research papers on the subject of the review."[6] But such well-meaning suggestions are impractical, too, given the number of manuscripts submitted annually to most journals. Tens of thousands of serials are published worldwide every year, each receiving hundreds and sometimes thousands of submissions.[7] Checking the credentials of all these authors would be such "a vast and embarrassing business" that, as one editorial writer points out, "editors would seem to have little choice but to trust the integrity of their contributors and the astuteness of their referees."[8] Even if authors were required to submit notarized resumes along with their manuscripts, an unscrupulous author might forge the notarization.

Conventions differ among disciplines and institutional settings, not only in notation styles but also in who may be listed as an article's "author" or included as coauthor.[9] Responding to confusion about this situation in biomedical science, Arnold Relman proposed that, to qualify as an equal coauthor, one must fulfill at least two of the following criteria: "conception of idea and design of experiment," "actual execution of experiment" or "hands-on lab work," "analysis and interpretation of data," and "actual writing of manuscript."[10] Even these standards differ according to employer. For example, in industry, authority in a research project is more segmented than in university-based research, but credit may go to the team as a whole. In industry, manuscripts may be composed by professional writers, based on a scientist's notes, and yet the technical writer is rarely listed as a coauthor nor is the writer's work acknowledged in print. These differences emphasize the need for sensitivity to variations in workplace requirements and cultures in determining publishing standards for authorship, from order of listing to inclusion.

Who's on First?

In Carl Djerassi's novel about scientific fraud, *Cantor's Dilemma* (1989), a scientist confesses that early in her career she changed her last name from "Yardley" to "Ardley" so that she might gain alphabetical advantage on coauthored manuscripts.[11] This effort later proves to be wasted when she is trained by a mentor who believes that senior scientists should always be listed after their students and Ardley adopts the same policy in her own lab. Tongue thrust firmly in his cheek, Djerassi uses the novelist's platform to explore one of the most sensitive indicators of scientific standing: who's on first?

In recent years, there has been a dramatic rise in the proportion of articles that are jointly authored.[12] The increase has been most pronounced in physics and the biomedical sciences. In the mid-1970s, one study showed, the rise in multiple authors appeared mostly in the biomedical journals; by 1981, the percentage of articles with multiple authors in an astrophysics journal was 67 percent and, in a medical journal, 98 percent.[13] In 1930, articles in *Lancet* had an average of 1.3 authors, and articles in *New England Journal of Medicine*, an average of 1.2 authors; by 1975, the *Lancet* average had risen to 4.3 authors and *New England Journal of Medicine* to 4.2 authors; by 1981, the *New England Journal of Medicine* average was 5.[14] By 1980, an overall index of the number of authors per article (for all fields), as calculated by the Institute of Scientific Information from its data base of journals, had climbed to 2.58 (up from 1.67 in 1960).[15]

The late Derek J. de Solla Price, speculating about why multiple authorship appeared to be on the rise, once suggested that it represented a "cheap way of increasing apparent productivity."[16] Now, even humor offers little consolation. Scientists bemoan the decline of single-authored papers; in some fields, teams routinely exceed a dozen or more collaborators per project.[17] The resulting fractionalized credit leads to new pressure to publish multiple versions of articles describing the same project, each with a different author listed first, a phenomenon labeled the "least publishable unit" syndrome.

To some, this phenomenon at first seemed laughable; annual awards were proposed for the scientific article with the longest list of coauthors. In high-energy physics, where teamwork is a necessity, a journal article in the late 1980s listed 130 collaborating scientist "authors" in a Stanford group, whereas an article contributed by a rival group at the European facility at CERN described four separate research teams, each involving over 200 scientists.[18] The absurdly high numbers obscure more significant problems. How, for example, could the collaborators on these papers even begin to agree on an order of listing that was not arbitrary or alphabetical? Because individuals routinely move into and off of large projects, whose name should be included, and whose omitted when publication is eventually achieved?

As the average number of coauthors increases, the overall order of listing seems more important; but Harriet Zuckerman's research shows just how arbitrary that order can be.[19] Her study of publications by Nobel laureates demonstrates that conventions of alphabetical or reverse-alphabetical schemes sometimes obscure the true "first author"—that is, the person who takes the greatest responsibility and hence should claim the greatest share of credit. Zuckerman explains that a form of noblesse oblige kicks in—the more eminent the senior scientist, the more "gracious" he or she might be over first authorship, sometimes allowing a junior coauthor to be listed first, even though everyone in the field refers to the work as that of the senior author. And, of course, it is the latter perception that enhances reputations.

Not all senior scientists are so gracious. The practice in some collaborative groups in the biomedical sciences is that the first author is "the person who writes the paper"; in many other groups, however, there is a "written policy . . . that the director of the group determines who will be listed as an author."[20] Senior scientists have been known to demand initial listing on articles in return for having allowed a postdoctoral or junior fellow to work in their laboratories.[21]

The reluctance of younger scientists to challenge these practices may be changing. In a dispute over how authors would be listed on a multi-authored, multivolume psychiatric text, the coauthor who had coordinated most of the editing sued the publisher for refusing to place his name first; he won the case on grounds of contract violation.[22] News articles about the episode drew angry letters to *Science*: "The credit goes mostly to those who are already prominent, and the labor is done by those who are not."[23] Or, as rephrased by one publishing executive: ". . . blood, sweat, and tears don't get you top billing."[24] The court, in this case at least, disagreed.

The order of authorship matters especially at the beginning of a career. With top billing, the young scientist will receive more credit for the work because colleagues tend to award more respect to first authors and because citations distinguish between primary and secondary authorship. Many fraud cases in the past decade have shown that the order of listing can also establish a record of who claims or should bear responsibility for an article.

Echoing the sentiments of many other scientists, a former president of the Institute of Medicine once suggested that all editors require authors to specify "who is responsible for what parts of the paper."[25] This practice has since been adopted by some major journals and it merits consideration by all as the number of authorship disputes increases.

Who's Along for the Ride?

Even more acrimonious debate surrounds the issue of "noncontributing" or "honorary" coauthors. As journalists pounced on the reports that a mere conversation in an elevator had sometimes been enough to earn coauthorship, the controversy over "free riders" grew.[26] Some journal editors began to report a growing number of requests from nonparticipating coauthors for removal of "gratuitously added names."[27]

Once the problem was identified, however, it was hard to know what to do. How can one separate fairly the legitimate contribution from the illegitimate? Who should police authorship practices? And who should punish violations? In many cases, even detecting who is or is not a fictitious coauthor is difficult. Cyril Burt, for example, apparently "created" quite plausible coauthors out of whole cloth; moreover, Burt was so prominent that questioning his coauthors would have seemed like questioning Burt himself.

Perhaps the most sensitive aspect of this problem involves the relationship of senior professors, mentors, laboratory chiefs, and thesis advisors to their students and junior co-workers.[28] Should "pupils" and "masters" receive equal credit for work that the pupil did in the master's laboratory? The goal of apprenticeship may indeed be training and education; but the senior scientist, also exploited as a student, may regard honorary authorship as a traditional part of the entitlement of senior status. In the Robert Gullis case in England, the professor assigned to supervise Gullis's doctoral research was listed as coauthor on seven of the eleven papers withdrawn; even though Gullis later admitted full responsibility, the senior professor was drawn into the case because of his coauthorship.[29] John Darsee frequently added colleagues as coauthors and yet, when the articles first appeared, none insisted that their names be removed. "It is routine courtesy for young researchers to get their seniors' signatures no matter how invisible their input," one journalist concluded, and "some lab chiefs insist [that] all work leaving their lab bear their name."[30] In reaction to such episodes, *New England Journal of Medicine* began requiring in 1983 that all coauthors' signatures be obtained prior to publication, a procedure now being adopted throughout scholarly publishing.

Although the central violation in the Robert Slutsky case at University of California, San Diego (UCSD) was his fabrication of data, authorship and coauthorship also became issues as that case progressed.[31] Questions were first raised about Slutsky's work during the review for his promotion to associate professor of radiology. Faculty members on the review committee just happened to read two of his papers side by side and to notice certain technical discrepancies.[32] Formal charges were brought in April 1985 and an investigating committee report issued in September 1986 stated unequivocally that Slutsky had fabricated research data.

Seven senior faculty members and dozens of young researchers (many of them postdoctoral fellows who had simply "passed through" the laboratory) were named as coauthors of articles that the university committee declared to be "questionable" or "fraudulent."[33] The university committee condemned the practice of honorary coauthorship in its report, calling it "a mockery of authorship." The committee's indictment was scathing:

> Some colleagues and trainees were either flattered or embarrassed to have their names put on papers to which they had not contributed, and many lacked the authority or willpower to resist this practice. On the other hand, some faculty [members] expected that their names be used even though they had provided only facilities for a project, without substantive contribution to, or knowledge of, the validity of the work.[34]

According to the UCSD committee, Slutsky used his coauthors to disguise the extent of his fabrications. At peak production, he was "creating" about one paper every ten days (in six and a half years, he submitted 161 manuscripts); by adding other names, he could make the output appear more plausible and possibly delay questioning.[35]

Even more disturbing to some committee members was the coauthors' apparent laissez-faire attitude toward the publishing process. Nearly all of Slutsky's co-workers, the committee wrote, left to him the task of "communicating with the journals and revising manuscripts" because he was eager, willing, and apparently effective.[36] For those articles on which Slutsky was listed as first author, none of the coauthors could recall having seen (or, by implication, having asked and been refused the opportunity to see) the written referees' comments, a practice the committee condemned as "an abrogation of co-authorship responsibility."[37] As a remedy, the committee recommended that journals require signed approval by all coauthors, the practice that *NEJM* had endorsed three years before. And yet this practice would still offer little defense against the person truly resolved to commit fraud. Slutsky had apparently forged some of the coauthors' signatures on copyright transfer permissions forms.[38]

Young postdoctoral fellows and residents whose names had been added to the fraudulent articles reported that they initially had thought Slutsky was doing them a favor: he had "given" them a publication.[39] But this gift horse had a short unhappy life. Many coauthors of Slutsky (as well as those of Stephen Breuning and others accused of fraud) since report that their involvement, no matter how innocent, has harmed them, because potential employers tend to treat the association as a stain on the coauthors' records.

To some people, listing a noncontributing coauthor perhaps seems worth the risk, or even standard practice in the field.[40] Association with a well-known senior person usually enhances the career of an unknown. In biomedical research, listing serves as a common gratuity for those who share data or supply biological materials (e.g., virus strains). The danger of these practices is that it lessens the responsibility coauthors may feel for the accuracy and quality of a manuscript; it can also lessen the opportunity coauthors have for checking a manuscript. As credit is parceled out, coauthors sometimes let one energetic team member shoulder all the work. When that happens, it can seem unfriendly or ill-mannered to recheck that colleague's work. Moreover, the more authors there are, the less qualified each becomes to judge an article's accuracy as a whole.[41]

These practices, regrettable as they are, evolved from a centuries-old practice of giving more credit to a master than to an apprentice, in all sorts of activities. For example, Rembrandt, Rubens, and other great painters maintained studios in which their apprentices did "much of the essential brushwork" while the master "acted as manager, guiding and polishing."[42] And until the twentieth century, art history accepted an "idea of genius," which assumed that because the master directed and guided a work, he should receive the credit.[43] Now, art historians are virtually dragging apprentices "out from under the Master's enormous shadow."[44] Connoisseurs explore the "personality" of each artist, and attempt to connect hitherto ghostly figures to the works they actually created.[45] Some art historians believe that the discovery of extensive but hitherto unacknowledged collaboration should also automatically raise questions about a painting's authenticity.[46] Whatever the decision on authenticity, however, effort is directed toward assigning paintings to the "correct" artist, even to the point of preferring to assign a painting to a younger, less well-known artist than to leave it vaguely attributed to "School of Rembrandt" or "Rembrandt's Studio."

One can adopt a parallel analysis for science which divides or classifies scientific manuscripts into comparable categories according to the degree of involvement of their authors.[47] At one end are those manuscripts written by and reporting on research performed by a single scientist. Then there are coauthored works for which one author designed the research, proposed the theory or closely supervised the research, but did not actually conduct the research or write the article. Finally, at the other end of this continuum are manuscripts written by junior researchers who conduct their work in the laboratory (or as part of the research project) of a senior scientist (who did not participate in the actual data collection described, or in its interpretation, but whose name is listed as a coauthor). Such distinctions help, in the case of science, to bring the "real" authors into the limelight, as the art historians attempt to do with apprentice painters.

Many scientists defend the practice of honorary or "gift" authorship— especially that awarded to senior scientists—by claiming that it is "standard fare." In fact, the codes of ethics of many scientific associations clearly prohibit taking *or* awarding unearned credit. The "Ethical Principles for Psychologists," adopted in 1981 by the American Psychological Association, state that psychologists should "take credit only for work they have actually done" (Principle 1, Part a). The code also specifies the difference between "major" and "minor" contributions for the purposes of assigning credit:

> Publication credit is assigned to those who have contributed to a publication in proportion to their professional contributions. Major contributions of a professional character made by several persons to a common project are recognized by joint authorship, with the individual who made the principal contribution listed first. Minor contributions of a professional character and extensive clerical or similar nonprofessional assistance may be acknowledged in footnotes or in an introductory statement.[48]

The Code of Ethics first adopted by the American Institute of Chemists in 1923 states in its 1983 revision that it is the duty of the chemist "to ensure that credit for technical work is given to its actual authors." Similar provisions in other codes emphasize the responsibility of an author to assign *proper* credit. International Committee of Medical Journal Editors (ICMJE) standards require that "a paper with corporate (collective) authorship must specify the key persons responsible for the article" and recognize others separately, perhaps in an "Acknowledgments" section. ICMJE also notes that editors "may require authors to justify the assignment of authorship," a policy increasingly adopted throughout scientific publishing.[49] The more extreme suggested requirements have been made in biomedical publishing; *NEJM* editor Arnold Relman, for example, believes that all coauthors of a manuscript should sign a statement attesting that they "actually had a hand in the research and are prepared to take responsibility for it."[50] *Annals of Internal Medicine* "requires each author to sign a form acknowledging that he has seen the final version of the paper and agrees to its publication."[51]

Even with the establishment of guidelines and policies at the journal level, change will only come if the research community fosters an environment that regards scientific credit as earned not bestowed. As attorney Barbara Mishkin has phrased it in testimony before Congress:

> Gift authorship, which means adding a senior scientist's name to a publication in order to increase its credibility in the marketplace, is like false advertising. It is promoting a product as one produced by a well-known and respected scientist when it was not.[52]

Who Makes Up the List?

Eugene Garfield believes that editors and publishers need not stand by helplessly; they already "have the authority to establish guidelines for authorship."[53] Without cooperation among the groups, however, guidelines are likely to be inconsistent and unworkable, because approaches can be so different. Some scientific journals require an alphabetical rather than status listing of authors. Sometimes long-term collaborators alternate the order on a group of articles. Many editors advocate, as standard format, initially listing the person who did the work and then, as last author, the person who supervised; authors would be required to attest to the accuracy of this order.[54]

Ideally, both authorship and order of listing should reflect "*scientific* contributions, most notably in the form of conceptualization, experimental design, analysis, interpretation, and presentation of work," Roger Croll suggests.[55] Authorship, he continues, should not just be "a reward for long hours of work" or for helping to write up the results, and credit should not be tied to status. He adds: "It is immaterial whether a collaborator is receiving academic credit, an hourly wage, a commission, or a salary in exchange for work"; "[a]llocation of authorship by any other method serves only to cheapen the credit given" to those listed.[56] Croll also believes that guidelines should nevertheless be voluntary because only individual workgroups, laboratories, or employers can apply and enforce compulsory rules. Scientific professional associations might encourage their members to adopt certain guidelines, of course, but no association or journal would have the means to monitor compliance.

WHO'S RESPONSIBLE FOR CONTENT?

Authors

Discussion of how credit is assigned, and to whom, eventually must focus on the bottom line—accountability. Who takes ultimate responsibility for the content? When authors carve up the credit, everyone's at the table. Taking responsibility then seems part of the reward. When suspicions of wrongdoing are raised, however, coauthors tend to disappear.

The disparity between how single authorship and collaborative co-authorship are treated is quite striking. "Most members of the scientific community equate authorship directly with credit and responsibility for an article," Croll writes.[57] But time and time again, dozens of coauthors whose names had been listed casually were caught up in fraud investigations and were defended as "innocent" parties who should be relieved of responsibility for false publications.

In many ways, publishing policies have tended to reflect the laissez-faire attitudes toward authorship which scientists expressed in the 1970s and 1980s. It became standard practice, in the name of efficiency, to allow one coauthor alone to handle submission and even revision of an article. A simple change could help to prevent problems arising from this practice. If journals require each coauthor to attest to an article's originality, then that effectively distributes the burden of responsibility. *Science*, for example, states in its "Instructions to Contributors": "By submitting a manuscript, an author accepts the responsibility that all those listed as authors of a work have agreed to be so listed, have seen and approved the manuscript, and are responsible for its content."[58] To confirm this agreement, *Science* requires a signed copyright form from every author listed.

In the social and behavioral sciences, joint efforts tend to involve interpreting interview data or clinical or field observations, for example, as part of a larger research project. In fields like geology or paleontology, where one's coauthors are likely to be colleagues who have shared their samples from a remote site, innocent collaborators are even more likely to be drawn into a scandal when the sites or the specimens prove false. Just such a dispute arose in paleontology in the 1980s, as prominent researchers in the United States, Europe, India, and Australia who had coauthored papers with Indian geologist Vishwa Jit Gupta found themselves caught up in an international scandal. Since the 1960s, Gupta had supplied his collaborators with ammonoids and other fossils that he claimed to have found in a remote site in the Himalayas.[59] In a coauthored paper, the paleontologist collaborator would describe the fossil and Gupta would describe the site. Because of the political problems in the region, independent confirmation was difficult and no one had reason to mistrust the well-known Gupta.

In 1987, Australian paleontologist John Talent began to accuse Gupta of lying about the fossils' origins—the specimens did not come from the Himalayas, Talent alleged, but from places like Morocco and may have been purchased from dealers. In 1989, Talent published a long paper in *Nature*, along with papers by other of Gupta's former collaborators, providing additional evidence.[60] Talent charged that the coauthors of more than three hundred articles had engaged in "carefree coauthorship" because they should have investigated the reliability of the data, they should have checked for internal consistency, "and, at the very least, they should have insisted that precise locality and stratigraphic information were given."[61] "Having had a sloppy approach to the primary facts," Talent wrote, they must now "take mutual responsibility for the defective products." Some of the Gupta collaborators, however, defended their actions by explaining that they had believed they were "helping the guy out." Science "is supposed to work on trust," some argued; they were simply "not trained to think in terms of dishonesty in science." Others admitted to being insufficiently "wary."[62] Gupta has vigorously denied the allegations.[63]

In a letter to *Science*, one of the coauthors disputed the accusation that the collaborators bore major responsibility; if fraud has occurred, then the coauthors are "the first victims," he wrote.[64] How can *they* be considered "sloppy," for they had every reason to believe the specimen fossils had come from the region described. He recounts in the letter how, when he was first shown some of the ammonoids, he recognized the similarity to Moroccan fossils; when Gupta continued to claim that he had "personally recovered" the fossils in the Himalayas, "I may have acted, perhaps, in too much of good faith, but certainly not with technical carelessness." In a rejoinder to this letter, Talent refused to consider the coauthors either "unwitting participants" or "victims." It's a matter of choice either way: "Few scientists have the inclination to 'blow the whistle' on fellow scientists' obviously spurious data"; instead, "they choose to ignore such problems," just as Talent did for sixteen years after he and some of his colleagues first suspected the reliability of the samples.[65]

Both the criticism made by Talent and the explanation of the coauthors ring true—and that contradiction exemplifies the dilemma for all long-distance collaborators. How much to trust and how much to voice skepticism? To Talent, the fact that so many scientists were apparently misled in this case just ironically confirms "the trust that most scientists have in one another."[66]

Condemnation of errant coauthors has been inconsistent, even when wrongdoing is proven or admitted. Few formal guidelines exist and the problem is actually a relatively new one. In a few cases, coauthors have been judged, at least initially, to bear full responsibility for an article later shown to be based on misrepresented or fabricated data. In other cases, they have been excused. An exchange between Stanford University official David Korn and representatives of the National Institute of Mental Health (NIMH) demonstrates how divergent the most sincere interpretations of the same situation can be. The NIMH investigator reportedly found it unrealistic to expect that the coauthors involved in the university case could have detected the problems identified later; but Korn disagreed. He replied: "I do not think that coauthors are blithely free to reap all the credits of putting their names on coauthored papers, and then disclaim any responsibility for problems."[67]

Derek de Solla Price expressed similar concerns in 1981 when he advised that "in the absence of evidence to the contrary," credit and citations "must be equally divided" among all coauthors. "We really ought to recognize as an ethical matter that each author in return for sharing the support also shares responsibility and credit," he added.[68] One hidden factor in coauthorship disputes may be the changing "nature of ownership of data, ideas or methods," Patricia Woolf explains, if the person who "owns" the data is also perceived as responsible for its accuracy.[69] If, as Woolf phrases it, "the notion of ownership has no meaning until the ideas are shared," then that same sharing could be construed to disperse responsibility. Research by June Price Tangney shows that, in fact, most scientists seem to regard the responsibility for data as their own, even after publication—though half the researchers she surveyed did state that, if they later had any questions about their own data, they might leave an article "standing unchallenged" in the literature while they attempted to replicate their results.[70] This timidity may result from publicity to recent arguments. Even to notify colleagues that one is reconsidering data might be construed as an admission (or accusation) of something far more sinister.

In part, the trust that scientists place in their coauthors reflects simple practicality. Every day we rely on many different sources of information, from weather reports on the radio to the data we use in our work. To verify every statistic, every interpretation in the things we read, would be impossible. Citations give pointers back to some materials, and a conscientious (or suspicious) reader may check the cited work, but few readers do this.[71]

Once a writer suspects errors or fraud in information he or she had innocently relied on, what should be done? Certainly no one should continue to use or cite data known to be fraudulent or inaccurate. But should the literature be purged of all articles that innocently relied on faked or forged data? For example, what should be done by the paleontologists whose theories relied on the authenticity of the Piltdown skull? Assumptions built on false ground must be reconsidered and possibly retracted; unresolved issues, like varying interpretations of data, should be described explicitly in print.[72]

Journals and Editors

Editors and publishers assume considerable responsibility for safeguarding the integrity of authorship, but there is little consensus on whether they should exercise these responsibilities passively or actively. At present, the most aggressive approach is to require every author and coauthor to sign a form consenting to the version published, but to pursue matters no further. Solutions such as including a section called "Attributions" in which each author lists exactly who did what (e.g., "Smith recorded the data, Jones analyzed it . . .") offer practical alternatives and are relatively easy to implement.[73] Other suggestions smack of confrontation, such as questioning "the presence of a senior scientist's name, and certainly an administrator's name, on a paper."[74] Perhaps editors do have the responsibility to deny publication until the name of a "noncontributing coauthor" is withdrawn from a manuscript,[75] but how could any journal confirm the status of every coauthorship? Here, again, defining the standards and establishing a climate of trust represent first steps. When a manuscript is submitted for consideration, authors can be required to attest that they meet the criteria set by the journal or else, as *Science* notes in its guidelines, explain why exceptions should be made.

DECEPTION, BETRAYAL, AND TRUST

In his book *Deceivers and Deceived*, Richard H. Blum sheds some light on why the issue of "authenticity of authorship" elicits such strong reactions from scientific editors and publishers. Blum studied confidence men, embezzlers, smugglers, and similar hucksters who assess the likelihood of establishing a relationship of trust with a victim and then, once it is achieved, who intentionally violate it.[76] In attempting to understand such people, Blum writes, the sociologist must look beyond specific acts and examine "a mix of the history and circumstances of the person, his groups, his situation, the politics of his times, the relationship between victim and deceiver, the psychodynamics of them both and, in addition, the morality and predilections of the observer."[77]

The deception that occurs when an author lies about authorship, either through plagiarism, false listing, or forgery, tends to be especially disturbing to editors because the editor presumes establishment of a "relationship of trust" and therefore feels betrayed when he or she discovers that the relationship "has not been what it ought, morally, to be."[78] At worst, the editor feels both guilty and victimized. Thomas Mallon, in his study of plagiarists, quotes an editor who unwittingly published an article containing plagiarism: "As editors . . . we ended up feeling tremendously guilty. It's, you know, not unlike rape . . . The victim feels as bad, as guilty, as the victimizer."[79] As Mallon rightly observes, the comparison is odious but nevertheless reflects accurately the depth of emotion elicited by deception. "At best," Blum writes, "the events judged to be trust violations have in common the judgment by someone that someone else should not have acted, given the confidences involved, as he did. A rule has been broken, a wrong has occurred, even though the rule may never have been written or even discussed."[80] An editor, immersed in his or her own scientific or scholarly work, assumes that each author subscribes to the basic rules of honesty about authorship found throughout publishing (or throughout science).[81] When an editor discovers otherwise, then the editor's own faith in the system may be temporarily shaken.

The *perceptions* of the person deceived are thus as critical as the actual deception. In human relationships that require trust and the exchange of confidences, such as the editor-author or editor-referee relationships, the parties have implicitly engaged "not to reveal or exploit their knowledge, position or advantages to bring harm to the other" or to bring harm to third-party partners to the relationship.[82] Blum demonstrates what most editors know intuitively—that is, the "giver of confidences is vulnerable because the deceiver appears to have accepted the usually unstated obligation to protect the self-interest of the other."[83]

Deception from an "insider," Blum notes, is all the more perfidious because it jeopardizes a certain core of idealism as well as undermines one's faith in one's own judgment.[84] The author, from the editor's perspective, appears to be an insider because he or she is usually also a reader, may have served or will serve as a referee, and is certainly a member of the research community served by the journal. If the author is not really the author, not really responsible for the article's success or failure, then the entire relationship collapses. Authorship appears to be a sham and, like the Cheshire cat, disappears before our eyes in a fog of dishonesty and mocking deceit.

What can each participant in the publishing process do to change existing problems and to prevent new ones? Journal editors and publishers have a primary responsibility to take the lead on this issue. Strong editorial statements about who does or does not qualify for listing as a coauthor can help to avoid disputes. Whether requiring validation of the authenticity of coauthors will deter those who are intent on deception remains to be seen, but it represents a positive step. Formal requirements help to define expectations and encourage honesty. As the editors of *JAMA* wrote when announcing a new policy of requiring authors to validate their responsibility for their articles, the "small additional bother . . . is designed to protect all of us from the shadow that has fallen over the scientific and medical communities."[85]

Discussions about authorship and accountability must, as Donald Kennedy has emphasized, begin within the individual project or laboratory, and dynamic change must start within the universities where scientists are trained and where the work is done, not just in the funding agencies. As a matter of principle, authorship should truthfully reflect who did the work and who accepts responsibility for it. Individual laboratories or work groups should set policies for allocation of credit and responsibility *before* a project begins, not when the manuscript is completed. Ideally, these issues should even be clarified before employment of a junior scientist or postdoctoral fellow, and discussions about credit should be initiated by supervisors, as a matter of good practice.[86]

Just as with other government regulations intended to address past abuses by preventing future problems, a secondary effect of tightened federal rules for authorship would be to heighten all scientists' sensitivities to issues of accountability. But it would be far better to change the ethical climate, rather than rely on punitive rulemaking to frighten potential wrongdoers into good behavior. The issue is absolutely clear. Who did what and how much? Answering those questions early on—and continuing to ask them as projects change—can help to prevent disputes or embarrassment later. The young coauthors of Darsee, Slutsky, and Breuning discovered too late that, if someone lies about your participation in a project and adds your name to a manuscript, they do you no favor. In such a circumstance, asking that your name be removed may be awkward, but discomfort is a small price to pay for integrity.

5   Decision Making
Editors and Referees

The eye sleeps, until the spirit awakes it with a question.
—Max Friedländer, on art history[1]

Maxwell Perkins was perhaps the consummate literary editor of the early twentieth century.[2] He discovered wonderful writers like F. Scott Fitzgerald, Ernest Hemingway, and Thomas Wolfe, shaped their prose, and displayed their work to new audiences.[3] But unlike the stereotypical literary editor, Perkins was no bookish grammarian, chained to his desk and unconcerned about a text's market. He clearly examined each new work with an eye toward its commercial appeal; he knew "what to publish, how to get it, and what to do to help it achieve the largest readership."[4] "Few editors before him had done so much work on manuscripts," his biographer concludes, yet "he was always faithful to his credo, 'The book belongs to the author.'"[5]

This attitude, that author and editor share a common goal of creating the best possible manuscript, infuses scientific publishing just as it does the literary world—in direct contradiction to the common perception that peer review alone rules editorial choice like some sort of routinized voting scheme. Many journal editors do sometimes allow referees' advice to overwhelm their editorial best judgment; they may acquiesce to a positive consensus of referees, even on manuscripts they regard as dull, uninspiring, or forgettable. Too timid? Perhaps. But these editors know that to solicit and then to ignore referees' advice altogether is to treat the review system as a sham. In truth, the journal editor usually has both a much stronger and much weaker position than conventionally assumed. The power to influence content lies not so much during the stage after review but much further back, in the process itself. And within that process lurks the problem of fraud and how to detect it.

A chief editor controls the review system because he or she sets its procedures and standards and appoints the referees; the editor's choices, however, are influenced by the journal's readership because *it* ultimately validates (accepts) or repudiates a journal's content. In addition, no editorship is static; each new editor changes a journal's editorial agenda and, in so doing, adjudicates among intellectual and political coalitions within the field. Editors, in the words of the late A. Bartlett Giamatti, "control the process in a fundamental way; they decide what will and will not enter the system."[6] Giamatti thereby articulated a sociological truism: editors most influence the eventual content by acting as the intellectual "gatekeepers" who monitor initial as well as final manuscript selection; they possess power because they can deny admission to some but admit others.[7]

In explaining the importance of this gatekeeping function, university publisher Jack Goellner cites two extreme suggestions for dispensing with the editorial function, ones commonly proposed by academics unhappy with the current system. The first type of person argues against any editorial filtering at all—instead, he explains, *everything* should be published. Of course, the resulting glut of information affects this academic very little because he just then employs assistants to screen the literature for him. The other type advocates eliminating conventional peer review and editorial structures "in favor of a licensing system" for electronic journals to which only a "hand-picked cadre of established scholars" would be allowed to submit articles. The flawed assumption in the latter case, Goellner notes, "is that whatever is written by those worthies will be worth reading by the rest of us."[8]

These choices, between publishing all or choosing some, are at the heart of the peculiar relationship between authors and editors. It is a relationship that can be simultaneously trusting and antagonistic because editorial decision making opposes two irreconcilable interests: no author revels in rejection and no editor enjoys publishing imperfect manuscripts. John Franklin Jameson, the great founding editor of *American Historical Review* (*AHR*), was apparently determined to establish "a standard of workmanship" in historical scholarship and writing.[9] Jameson ruled the journal with a strong pen, asserting that *AHR* must first serve the best interests of its readers and then those of its writers, but his criteria embodied his own biases about the study of American history; because he was the ultimate arbiter of those interests, alternative interpretations made few appearances in the *AHR*.[10]

Similar conflict stems from the emotion that surrounds the act of writing. Manuscripts, whether they are works of poetry or paleontology, possess a special, highly charged emotional status in their author's lives. Our words represent the products of our minds; we treat our books like children that pass through phases of conception, gestation, birth, socialization, and sometimes death. However, to edit is to change. The very nature of the editor's job requires standing over someone else's intellectual progeny, wielding a blue pencil like a knife. This reality helps to explain why the popular images of editors are so negative. The mass media frequently portray editors as irascible or grouchy characters, as in the movie *Citizen Kane*, the play *The Front Page*, or the television series "Lou Grant." In the cartoon "Shoe," a reporter attempts to solve a crossword puzzle and asks his boss: "What's a six-letter word meaning 'odious reptile'? . . . 'Editor' doesn't fit . . . I tried it." Rick Crandall quotes typical criticism of editors that characterizes them as "bloody-handed . . . axe murder[er]s . . . against the life work of others."[11] In truth, a better metaphor involves not violent human interaction but horticulture—sometimes one must graft, prune, or transplant in order to produce a better flower or fruit, but then one feeds and waters the plant, pats the earth to provide a solid foundation for the roots, stands back to appreciate the resulting beauty, and prays for good weather.

Jacques Barzun, in a sharp, humorous essay on the editorial relationship, explains the antagonism as systemic.[12] Authors, he writes, traditionally see editors as intervening in the conversation between author and reader, as destroying their prose and hence their ideas, as needlessly and recklessly altering their pristine texts.[13] In addition, the editor stands in for another traditional antagonist, the publisher. The opposition of authors and publishers "is worse than the natural one between buyer and seller, because it is complicated by pretensions and temperament on both sides," Barzun writes. "Authors, as everybody knows, are difficult—they are unreliable, arrogant, and grasping. But publishers are impossible—grasping, arrogant, and unreliable."[14]

Editors, too, cling to exaggerated self-images, tending to view themselves as "overworked and abused by authors" and (especially when they are researchers or professors in the same field) as providing an unappreciated service for the community. Crandall believes that some editors exploit their seeming altruism and sacrifice just to "avoid the responsibility of proving that they are doing a good job."[15] They treat authors as hungry for publicity and little deserving of the "privilege" of publication in their journal and use that as an excuse for neglecting their editorial responsibilities.

As with most things, the truth is somewhere in between. The ideal relationship between editors and authors contains mutual benefit and runs on mutual trust. Authors need editors and publishers to disseminate their work to the appropriate audience and in an appropriate and agreeable form. Editors need good manuscripts to enhance their journal's reputation. Publishers need manuscripts to fill the pages with content that subscribers want to buy. The editor of the *British Medical Journal*, Stephen Lock, believes that authors "are entitled to expect" that editors will treat articles promptly and courteously, and that they will also reconsider an article if it is turned down.[16] The complexity of these unarticulated feelings can make it difficult, however, to understand what's going on when an accusation of fraud has surfaced and the various participants in the publishing process are trying to decide what to do.

MANUSCRIPT PROCESSING

When I use the word "editor" in this book, as in the passages above, I intend it to encompass the entire editorial structure, because an editor accepts final responsibility for and frequently represents the journal not as an individual but "on behalf of the journal." Few journal editors really work alone. They are assisted in decision making, processing, and policy making by paid professional staff and an international social network of advisors, referees, and other editors. Board members appointed by sponsor or publisher may assist in policy decisions. Advisory board members and sub-editors (e.g., associate editors or consulting editors) may accept some responsibility for content but usually in an advisory role. No set rules govern editorial or administrative arrangements for journals. No one system stands out as "best."

In most operations, a paid professional editorial staff runs the office and manages manuscript flow, review processing, and copy editing, sometimes with the assistance of students who volunteer or receive course credit. At the publisher's central offices, paid professionals supervise production, deal with vendors (e.g., typesetters and printers), market and promote the journal, and manage all aspects of subscriptions, circulation, and sales. The size of an organization and the hierarchy of responsibilities vary from journal to journal. Such things as the number of auxiliary editors and their authority, the number of referees used, how referees are chosen (or by whom), and the percentage of manuscripts that undergo external review all determine the amount of professional, paid staff needed to support the editorial office, whereas circulation size usually determines the extent of production and management. Editorial and production processes may be housed together or be long-distance operations. Companies that publish dozens or hundreds of journals may consolidate production responsibilities and may negotiate blanket contracts with vendors who supply paper or typesetting for all their journals. Understanding the general administrative model for each particular journal, especially with regard to the lines of responsibility and authority and where mistakes may have occurred, is critical for assessing how a publication may respond to allegations and

Table 3. Typical Duties of a Chief Editor

• Determines the scope of the work of the journal.

• Plans and authorizes special issues or sections.

• Appoints and supervises the work of subeditors, staff, and advisory boards or committees.

• Selects peer reviewers (or supervises their selection) and receives their advice.

• Makes final decision (or ratifies decision) on acceptance and rejection of manuscripts.

• Certifies when issue is completed and ready to print.

• Negotiates legal agreements or other formal arrangements for the journal.

• Assumes some legal responsibility for content, according to contract with publisher or sponsor.

• Reports to the publisher (and/or advisory committee or sponsor).

---

investigations of wrongdoing. The organizational model determines not just *whether* an editor has the authority to cooperate but also how swiftly, efficiently, and comprehensively even requests for information may be fulfilled.

Table 3 lists typical duties and responsibilities for a chief editor.[17] Alternatives to the "chief editor" model include co-editors (who may divide equally or share responsibilities) and editorial committees (appointed for their intellectual reputation, technical expertise, and professional accomplishments not necessarily their editorial expertise).[18]

Commonly, as the content of a journal diversifies and the number of manuscript submissions increases, a chief editor will add associate editors who are specialists in new areas and who share in the decision making. For example, in one year, the journal *Environmental Science & Technology* appointed associate editors "to provide more expert coverage of the broad range of papers submitted"; a few years later, the internal editorial operations were reorganized once more to allow associate editors more responsibility for choosing referees and for communicating directly with authors about revisions. This arrangement replaced an earlier structure in which paid professional staff in the main office chose referees from a computerized data base.[19]

Ultimately, someone must decide exactly what content to publish, and usually the chief editor accepts the final responsibility. The model of an autocratic, buck-stops-here ruler has a long tradition. As written about Franz J. Ingelfinger, one of the most famous editors of *New England Journal of Medicine*, "Ingelfinger is hardly the type of editor who merely collates the results and prints according to majority vote. He is the ultimate authority, the autocrat."[20]

With success comes an ever-mounting pile of manuscripts, however, and common sense requires that editorial responsibility be dispersed or more layers of editorial review be created. At *Science*, for example, members of an Editorial Review Board, who are experts in fields relevant to the journal's content, screen the manuscripts, evaluating "the significance of the conclusions" on the assumption that those conclusions are defensible. According to former editor Philip H. Abelson, about 40 percent of incoming manuscripts are rejected at that stage. The remaining are divided among knowledgeable staff editors who then choose "panels" of peer reviewers, guide the manuscripts through the process, oversee the quality and thoroughness of the review process, and eventually present the decision for final consideration in an editorial staff meeting.[21]

At the *Journal of the American Medical Association (JAMA)*, all manuscripts are reviewed by the editor-in-chief or his deputy editor and then assigned to one of ten physician-editors.[22] Each physician-editor represents a set of medical specialty areas and has the "authority to consult, reject, or work with the authors for revision." The physician-editors initially review manuscripts according to a checklist of such criteria as originality of ideas and material, appropriateness of analysis, reasonableness of conclusions, and adequate data or evidence. They then assign it to referees. After external review of a manuscript, the physician-editor decides whether to reject it, to return it to the author for more revision, or to consult with other editors. Manuscripts are formally accepted by the editor or deputy editor, but the recommendations of the physician-editors are crucial. The tiered system, created to handle the thousands of manuscripts *JAMA* receives annually, assigns to each sub-editor expertise within a limited domain; but the senior editors of *JAMA* must be sufficiently conversant with the broad range of topics to assess the worthiness of the sub-editors' recommendations and to make a final decision. Both *Science* and *JAMA* represent the most complex type of decision-making system.

During this phase at any journal, editors adjudicate among their conflicting obligations to authors, subscribers, and referees. Do all authors have the right to have their articles externally reviewed? A journal that solicits external peer review for every manuscript, no matter how unsuitable it may seem, no matter what its obvious problems, quickly exhausts its pool of volunteer referees and depletes its office budget. What if a manuscript is, for example, obviously libelous or includes insulting racial stereotypes? What if the biases are apparent but not blatant? What if a manuscript is so poorly written that rejection by even the kindest, gentlest referee is inevitable? Compulsory refereeing may *seem* universally "fair" to authors but it also unfairly exploits the time of referees, editors, and staff.

Reliance on editorial review alone may not be the best alternative to compulsory review. Rick Crandall, citing research in sociology, asserts that the editorial process in general "has yet to demonstrate either reliability or validity"; most processes produce results that are inconsistent in quality.[23] Unevenness of review (e.g., no consensus on review procedures), multiple biases (e.g., institutional, gender, personal, and intellectual), provincialism, and breaches of confidentiality are all problems mentioned by other critics.[24] In practice, however, editors more often err in the other direction; they keep far too many manuscripts in the pool for far too long, out of reluctance to offend, out of timidity in the face of a "difficult" author (especially one who is a senior figure in the field), or out of an unrealistic attempt to seem "fair." Franz J. Ingelfinger once observed that *NEJM*'s editors tended to accept about 10 percent of manuscripts they should have rejected and reject "about 10 percent that we should have accepted."[25] No one wants to lose a hidden gem, of course; but manuscripts can go through so many rounds of refereeing that an editor may finally consent to publication out of exhaustion.

Staff education, experience, and training play equally important roles in the expediency and efficiency with which each journal processes manuscripts. An alert and well-trained staff can facilitate early detection of problem manuscripts. And professional copyeditors and indexers have confided to me that they have often detected serious factual errors (and even plagiarism) during their work with manuscripts, all of which they report to their employers but not all of which were corrected. Some observers cite the referee as the journals' "sole defense" against fraud, but a stronger defense includes an alert and well-trained staff.[26] They are often the only people who scrutinize *every* manuscript.

The editorial staff members' attitudes toward science are another crucial but neglected factor in how a journal might treat a case of unethical conduct. The former managing editor of *Environmental Science & Technology* once described her role as that of helping to maintain the "cathedral" of science:[27] "In our own small way, we are helping to advance science by insuring that *ES&T* continues to publish original and significant research."[28] She noted that, although she was a chemistry major in school, she had had little familiarity with how the peer review system operated until she began a career in publishing.

Strong belief in the importance of science and usefulness of editorial review can help to bolster staff morale and vigilance in the face of what is invariably an ever-mounting volume of manuscripts. Some journals now publish several thousand pages each year. The *Journal of Physical Chemistry* grew from 5,000 pages in 1982 to 7,000 pages in 1986.[29] Each manuscript published, however, is matched by many more not published, which the editorial staffs must still process and check. At over 40,000 research journals published around the world, one writer estimates, articles are submitted at a rate of one every 30 seconds (24 hours a day, 365 days a year).[30] The high-circulation, high-visibility biomedical journals receive thousands of manuscripts every year. *NEJM* reported receiving about 2,800 articles in 1977 and publishing about 12 percent of those. Although about 20 percent were rejected out of hand (usually because the subject of the article was inappropriate for the journal), about 2,000 referees were asked to review the rest.[31] In 1989, *NEJM* reported receiving about 3,600 unsolicited manuscripts annually, in addition to several hundred book reviews, editorials, and commissioned essays; they published 10–12 percent of the unsolicited manuscripts and "the great majority" of the solicited.[32] The amounts processed at other large biomedical journals are comparable. In 1983, *JAMA* published 704 articles, selected from among 3,600 submitted. In that year, the acceptance rate for all manuscripts was 19 percent; for unsolicited manuscripts alone, 13 percent. More than 3,000 individuals reviewed manuscripts for *JAMA* during 1983, but that figure did not include the individuals engaged in editorial and expert review of the thousands of manuscripts submitted or commissioned as letters, news, book reviews, obituaries, or announcements.[33]

Most journals receive hundreds, not thousands, of manuscripts each year. *ES&T*, a well-regarded journal in its field, reported processing 280 manuscripts in 1977, 350 in 1982, and 428 in 1990.[34] In 1987, the same journal reported that it took an average of 49 weeks to process a manuscript (14 weeks alone in the production stage), and it was revising its processing system.[35] Half of the processing time was taken up by peer review and/or authors' efforts at revision; but in the period 1981 to 1983, *ES&T* accepted 60 percent of the manuscripts with little or no revision required.[36]

Intense competition or dramatic advances in a field usually speeds up the flow of manuscripts. During the spring of 1987, *Physical Review Letters* received about a hundred manuscripts on superconductivity alone in the space of a few months. To cope with this unexpected amount, *Physical Review Letters* and *Physical Review B* established special "review panels of scientists who could render rapid judgments on the merits of each submission."[37] Authors at first used overnight express mail for delivering manuscripts; by April 1987, many were delivering manuscripts in person or by courier. By 1989, journals in other highly competitive fields reported receiving manuscripts by fax.

If a high proportion of the manuscripts submitted are eventually accepted, then the number of manuscripts at various stages of the system at any one time can be staggering. Combining data from two studies, *The Scientist* reported in 1988 that overall acceptance rates ran from under 20 percent in sociology and psychology to over 60 percent in physics, biology, and chemistry, and 90 percent in astronomy.[38] At the low end, even a small-circulation journal may process (and, it is assumed, check at least perfunctorily) several thousand manuscript pages each year for publication and, depending on acceptance rates, screen at least twice as much during the review stage; at the high end, a successful, high-circulation journal may process hundreds of thousands of manuscript pages annually.

When the number of submissions increases, the costs of processing manuscripts and running the internal and external review systems rise.[39] Separating out these costs for analysis is difficult because the tasks in an editorial office are not sequential. Each time a manuscript goes to or from a reviewer, for example, it must be tracked in the editorial office. One of the few published estimates is that of Arnold Relman, who in 1978 put the annual cost of the *New England Journal of Medicine* peer review system at six to seven person-years of referee time, two person-years of in-house time, and about $100,000 in office expenses.[40] Other editors advise that the review process for large journal operations can cost hundreds of thousands of dollars each year. Adding new checks would inevitably raise the cost of journal operations.

The concern about scientific fraud has already resulted in other increases in staff costs. Biomedical journals such as *JAMA* and *NEJM* now employ statisticians who supplement the referees' broad technical review.[41] When a journal pays for statistical review, it usually applies that expertise late in the review process, to articles that have already passed some initial round of scrutiny. Because the limiting factor is money, some publications depend more on expertise donated by university faculty or ill-paid graduate students, but even these costs can become prohibitive at times. The *Journal of American History* made the following remarkable announcement when it reassigned responsibilities in the editorial offices:

> To serve all our readers, we have cut back on a traditional service to our authors. No longer will we check the accuracy of every footnote citation to sources in the Indiana University Library. Instead, an assistant will spot-check the accuracy of some citations, and we will check generally for accuracy of transcription of titles and lengthy quotations. . . . The experience of the *Journal*

confirms the widely held observation that sloppiness in historical documentation is widespread. But the problem is much too broad, and the solution of checking every reference in one library is much too narrow and expensive. We hereby inform our readers that we are following the conclusion of three-quarters of them and of most other history journals that the accuracy of footnote citations is primarily the author's responsibility.[42]

Few journals have probably ever regarded "assurance of authenticity" as their primary responsibility, as expressed so conscientiously by the *Journal of American History*. Whatever the additional technical checks used, vigilance boosts the overall editorial expenses.

CONFIDENTIALITY OF THE EDITORIAL PROCESS

The degree of openness and secrecy present within a journal's review process appears unrelated to size, influence, or type of content. Some review processes are *private*, involving only the editor and/or editorial board and rarely any "external" referees. At other publications, review always includes external evaluators who know the identity of the author. Identification is a one-way street, however, for the author may not even receive verbatim accounts of the decision criteria, much less know who the reviewers are. At the other extreme are journals that espouse totally *public* procedures, all evaluations on the record and the author informed about the procedure and its outcome, as well as the names of the referees.

Neither practitioners nor analysts agree on which mix of openness and secrecy provides better protection to authors or a better review. All do agree that some confidentiality—at least while a manuscript is under consideration or in production—is recommended. For example, the Committee of Editors of Biochemical Journals, International Union of Biochemistry, explained as follows this recommendation in its 1979 Code of Ethics: "'All manuscripts received in the editorial office should be considered privileged communications, and be so identified.' A privileged communication may be defined as a confidential document not to be shown or described to anyone except to solicit assistance in reaching an editorial conclusion provided that this privileged status is made clear to the referee.'"[43] Privilege is normally extended to all editorial activities and correspondence between authors and editors.[44] In most circumstances, however, editorial confidentiality tends to be an ad hoc procedure.

On occasion, a journal has taken a strong stand to protect confidentiality at all costs. One dramatic example involved the ongoing political debate in the United States over the use of euthanasia for terminally ill patients. In an article published in *JAMA*, an anonymous author had confessed to having helped a terminally ill patient to die at her request; the author's intent was ostensibly to illuminate a controversial and emotionally charged ethical dilemma faced by many hospital physicians every day.[45] When the district attorney for the City of Chicago (and, subsequently, a grand jury) demanded that the journal's editor-in-chief, George D. Lundberg, reveal the author's name and turn over the editorial records for the article, Lundberg refused to cooperate; his publisher, the American Medical Association, which has its offices in Chicago, supported his decision. Lundberg claimed immunity from the government's request under the First Amendment and the Illinois Reporter's Privilege Act, arguing that the author had requested anonymity as a condition of publication. Lundberg's public statements about the case indicate that, in this particular case, his roles as editor/journalist and as physician came into sharp conflict, but he chose to be guided by the practices of the former because he had been acting in that capacity in attempting to promote debate on a crucial matter of professional practice.[46]

Lundberg's actions emphasize the importance of trust relationships between a journal and its authors. Similar confidentiality issues have been raised by a 1989 U.S. Court of Appeals decision pertaining to peer review. Ruling in *Arco Solar Inc. v. APS*, the court stated that a professional association may maintain the anonymity of journal peer reviewers in order to protect the integrity of the reviewing process and that such interests outweighed, in that case at least, the plaintiff's need for information. Both the Lundberg and the Arco cases have significant implications for how an editor may or may not be able to protect, say, a referee who accuses an author of plagiarism or data fabrication.

The confidentiality of journal peer review can inadvertently delay the investigation of fraud and misconduct if an editor suspects that any formal inquiry by a university or government agency might breach editorial secrecy. It's much easier just to reject the manuscript without comment. Those attitudes that lead editors to treat their own confidentiality policies with self-righteous cavalierness or to back away from confrontation stem from fuzzy definitions of expectations and obligations. It may be practical for journal editors both to have and to share secrets, to keep control over certain types of information in order to control how it is used, but ambiguous definitions promote a climate in which ad hoc policies seem acceptable.[47] Those editors who routinely breach review confidentiality, "using" private information to gain power or be perceived as powerful, should be condemned and those who chose to remain honorably taciturn should be supported. As Franz J. Ingelfinger wrote in 1974: "The ideal reviewing system is marked by an utter respect for confidentiality."[48] A fraud investigation tests these assumptions dramatically.

WHAT IS PEER REVIEW AND WHY BOTHER WITH IT?

The vast majority of attention paid to journals tends to focus on peer review, a term applied to technical scrutiny of manuscripts by experts. The use of peer review legitimates and authenticates scientific and scholarly journal articles in a way not duplicated in popular periodical publishing; it gives these texts authority and credibility. Furthermore, the emphasis on peer review reinforces a myth that says all scientific journals use rigorous expert review in selecting all content and that the peer review process operates according to certain universal, objective, and infallible procedures, standards, and goals. Quite the opposite is true, however. In some fields, the greater proportion of journals and monographs do receive expert evaluation prior to publication; yet, important, influential, high-quality works are published in the same fields without external evaluation.

In its ideal, expert peer review determines not the quality of the manuscript on some absolute scale but whether the report should be published in the particular journal and, if so, what changes will be needed before publication. Dozens of different criteria may be applied during this process. At the extremes, the decision-making process can take as little as hours or as long as years, but the outcome must always be either "yes" or "no."

To qualify as true peer review, a process must contain some possibility of rejection. Expert advice must not be sought simply to ratify foregone conclusions or to provide comments for revision. Independent evaluation must actually influence the editorial decision. This particular aspect of peer review is essential for understanding how research fraud affects a journal: peer review is neither uniform nor totally reliable nor intended as a fraud detection mechanism. Its principal goal—and perhaps what should be its only goal—is to evaluate manuscripts according to whether they should be accepted or rejected, not to determine their authenticity.

The peer review procedures so often touted in political settings as ensuring scientific authenticity, accountability, or authority are simply arbitrary creations and, like other human creations, they are fallible.[49] The characteristics of the research community that a journal serves, as well as that journal's own history, are far more likely to have shaped the extent and particulars of review procedures than any universal standards within the field. No set rules govern how, when, or by whom all journal peer review is conducted. At the level of individual journals or publishers, procedures conform to practical considerations such as financial and staff resources; they are affected by chance; and they are shaped by the values and experiences of every editor involved in each decision.

Journal editors tend to regard referees' reports as nonbinding; referees assume they have made a perfect assessment; and authors react positively or negatively, depending on the outcome. Most editors will not second-guess referees—especially if all reviews have been negative—but most refuse to treat reviewers' reports as the final word on a manuscript, for their recommendations must always be combined with such considerations as standards of length or format, writing quality, legal right to publish, and the topics a journal wants to emphasize.

Ideally, formal peer review procedures allocate scarce resources. They recommend whether a given manuscript should be published by a given journal; they do not choose the best article ever written on the topic. Peer review neither judges nor decides whether the manuscript is publishable anywhere else. Each editor can select only from the pool of manuscripts submitted and select only as many articles as can be published in the space available.[50] Writing in 1974, Franz J. Ingelfinger pointed out that editors are always in a "no-win" situation: to make a journal successful requires one to attract better and better manuscripts; but to do that, a journal must increase the total number of submissions and therefore increase the total number of manuscripts rejected.[51]

Some prestigious journals use external referees for just a tiny fraction of their manuscripts; others adjust the amount of external review to the author and topic. Members of the National Academy of Sciences, for example, have the privilege of publishing their own work in the Academy's *Proceedings* with no more than an informal review; articles by nonmembers must be transmitted via a member and receive anonymous review by two external referees.[52] The editors (almost always, in these cases, eminent scholars or scientists in their own right) make most decisions themselves.[53] Other publications obtain reviews from at least two referees per manuscript (or per round of reviewing) but may override recommendations if referees have recommended either far more or far fewer manuscripts than may be needed to fill the space available.[54] For some time, external refereeing has appeared "to be accepted as immemorial and inevitable as rain or cold," though it is neither.[55] Instead, the review system frequently serves as a way to validate editors' judgments by imparting a sense of thoroughness, clarity, objectivity, and uniformity even when those qualities do not exist.

When their reviews receive fair consideration by the editors, the diligence and expertise of the particular referees are keys to a journal's success. The individual referee anchors the review stage; John Ziman calls the referee the "lynchpin about which the whole business of Science is pivoted."[56] A referee's ability to criticize, the extent and depth of the referee's knowledge of the topic, and the overall alertness and diligence will determine the review's accuracy. The importance assigned to review seems all the more remarkable when one considers that most referees receive no compensation (or only a token honorarium) for their work.[57]

Considering the admitted failings, cumbersomeness, and cost of the peer review system, a skeptic might wonder why it survives. Why have journals not substituted some more efficient and reliable form of evaluation? Or why don't more editors just trust their own judgment? Or appoint special review committees? Part of the answer lies in the fact that peer review is driven as much by authors and readers as by editors. Authors seek publication in peer-reviewed journals because expert review provides the appearance of external and independent validation of their work.[58] "As producers, scientists want to have their work protected . . . validated, and then published to their peers . . . ," Stephen Lock observes.[59] There is also pressure from the readers. As consumers of journals, they want some assurance, without the effort of checking it themselves, that the work of their colleagues is authentic.[60] Peer review procedures are admittedly time-consuming, but alternatives such as standing committees might be even more so.

No two people tend to assess the peer review process or its goals in quite the same way. Drummond Rennie concludes that reviewers regard themselves as "unthanked, ignored, or reviled, [giving] freely of their energy and time, often reanalyzing, redirecting, and rewriting papers to improve their quality and to enhance, and sometimes save, the author's reputation, while guiding the editorial decision."[61] Authors, however, see referees as "biased nitpickers, impervious to originality, jealous of excellence and bent on delaying publication, breaking confidentiality, and plagiarizing ideas while solidifying prejudice and ensuring that the rich of the establishment get richer."[62] Reconciling the differences in these perspectives when disputes arise can pose one of an editor's most difficult problems.

Given the inflated perception of its value, it is remarkable how little we know about peer review.[63] Few comprehensive studies have examined the specific criteria used for decision making (or even the criteria that journals suggest to their referees beforehand); most have concentrated on case histories of single journals. How do referees apply standards? Do they do so consistently? Although it is fashionable to describe "the process as automatic and self-correcting," historians like Nathan Reingold point out that scientists' personal papers tend to disclose "a complex, human process."[64] If, as Harriet Zuckerman and Robert Merton conclude in their study of refereeing, the system "provides an institutional basis for the comparative reliability and cumulation of knowledge," then we must take into account the effect of what is clearly acknowledged to be a changing social system in science, where there is enhanced competition, exaggerated calculation of value, proposal pressure, and a sense that high principles are rarely rewarded.[65]

WHAT DO REFEREES DO?

Myriad complaints surround refereeing, including the frequent accusation that referees will recommend acceptance or rejection without reading the manuscript or that others only agree to review manuscripts that allow them to sneak a look at competitors' work or to guard their turf through criticism of potential rivals.[66] The pool of referees for a subfield, especially a new area, can be surprisingly small. If an editor judiciously attempts to avoid using the collaborators, colleagues, or competitors of an author as referees, the pool of available experts is even smaller.

As with other volunteer activities, exceptional performance as a reviewer attracts more requests for one's time. There is strong evidence that in specialty areas referees are chosen repeatedly from the same group. According to internal studies conducted by the journals, the "average" referee for the *American Journal of Public Health* (*AJPH*) also reviews for three to four different journals every year, and one-third of those who referee for the *British Medical Journal* review for other journals; *AJPH* surveyed a sample of 264 reviewers who, in 1987, reviewed for a median number of 3.6 journals.[67] Prior experience as a referee thus tends to enhance the chance of reviewing in the future, and editor Stephen Lock has further linked active participation as a referee to the level of activity as an author (or as editor of another journal). Authorship brings scholars and scientists to the attention of those who select referees, but referees may also be recommended by other referees or by journal advisors. It is common practice, for example, to ask referees some question like "Who else would you recommend as a reviewer on this manuscript?" These practices help to bring young scientists to an editor's attention, just as they also inadvertently perpetuate the "same old networks."

Whatever their motivations for agreeing to read a manuscript, referees provide an invaluable and unduplicated service because they evaluate a work apart from its context. Even if referees are given the name of the author of a manuscript, they rarely know other pertinent factors such as the manuscript's prior history, any preceding discussions with the editor, and the number of manuscripts on similar subjects that may also be pending. By examining a manuscript in isolation, the skeptical referee may see errors or incongruities that will otherwise escape attention.

Another serious criticism of journal peer review is that all referees do not apply the same criteria when evaluating manuscripts. Lack of consistency among referees could be considered as something that invalidates the system's usefulness, but it has been a flaw that many ignore. Exactly how precise should the journal's instructions to its referees be? Exhaustive "checklists" do not necessarily assure comprehensive review, whereas open-ended review does encourage healthy skepticism.

Journals generally rely on referees to evaluate manuscripts according to multiple criteria, some of which are unique to the type of research described, and some to the specific manuscript. The problem of how to balance ratings of pre-set criteria (e.g., quality of writing or use of supporting evidence) that apply to all works with the broad scrutiny of an individual work parallels the task of art criticism. In evaluating authenticity of a painting, for example, art historians use standard points of comparison like type of subject, paints, size of canvas, and brush technique for the presumed artist.[68] Experts initially consider how a work fits into the artist's conventional *oeuvre* (for example, can its subject matter, style, and size be associated with those of other paintings known to have been painted by the same artist during the same period).[69] Yet, sometimes a forger adapts unconventional techniques or attempts a subject unexplored but plausible for that artist. Sometimes a forger exploits incompleteness by "creating" a preliminary sketch for an authenticated painting. For art critics, the authentication process must include assessment of the context of what is known about all of an artist's work, even though a work must also be reviewed in isolation from that context.

In peer review of a scientific manuscript, the context drops away but also only for the moment. Referees are supposed to address the technical accuracy of details or statistics, to judge whether the work submitted appears on the whole to be correct, to assess whether the conclusions fit the data presented, and to assess the manuscript's readiness for publication (e.g., how little or how much work is required to ready it for publication). Referees may also be asked to advise on the scientific legitimacy and standing of a manuscript, on its originality (which requires comparison with work by others), and on the importance and relevance of its results. If the author's name is revealed, then referees may also compare the article to the author's other works.

With these requirements in mind, an unscrupulous author might craft a manuscript to fool those experts likely to review it. This phenomenon has not been systematically studied for science, but it is well-known in art. When a fake Franz Hals was presented to the critic Hofstede de Groot, for example, its acceptance was assured because the forger, knowing that de Groot was the person most likely to evaluate the painting, had tailored the painting to follow de Groot's own writings on Hals. The forger skillfully embodied the expert's "own style criteria" for authenticity.[70]

Aside from characteristics unique to each journal's style and format, for what types of characteristics do referees look? One guide to refereeing suggests three common factors: originality, soundness/validity, and significance.[71] Other analysts list such criteria as replicability, importance, competence, and intelligibility. At *JAMA*, referees are asked to "grade the manuscript's overall quality, estimate the priority of importance, recommend acceptance, revision, or rejection, and advise speed of processing should the manuscript be accepted."[72]

The crucial question for the study of fraud is, of course, whether referees can or should be expected to detect fabrication, forgery, or plagiarism. The level of skepticism stands as one barrier. The relationship between referee and author—even if anonymous for both parties—rests on the same type of trust as the relationship between editor and author. Arnold Relman writes:

> The peer reviewers start from the assumption that the author is telling you the truth when he tells you what he did and what he saw. . . . You don't know he isn't telling the truth unless what he says is inherently unreasonable or improbable and doesn't add up.[73]

Evaluation of a manuscript ideally works as a "partnership" between editor and referee, both employing similar high standards and both seeking to accept only articles that meet those standards. Editors must accept responsibility if they have been negligent and have failed to detect gross forgery or plagiarism; but few people blame referees as long as they have conscientiously executed their review and have examined a manuscript with due diligence and care.[74] Problems more often occur when referees themselves are slack and hide the fact from the editors.

It is possible now at some journals for a referee to request additional data (or more information on experimental controls), but such requests are still apparently rare. What if a request of this type results in a longer manuscript and hence conflicts with restrictions on space available? If appropriate supporting data are not supplied in (or with) a manuscript, who is responsible for insisting that they be provided? Unavoidable conflict may arise between accepting a manuscript and "impugning the integrity of the author."[75] Certainly, any such request will delay the decision process. How and whether this type of conflict is resolved relates directly to what all participants regard as the underlying goal of external peer review because peer review seeks to evaluate intellectual content, not deception. More often, referees are criticized for being "insufficiently critical,"[76] because like editors, they must maintain both trust and skepticism simultaneously.[77]

Failure of the system occurs when a referee has insufficient background to judge a work but is reluctant to admit it, or when evaluation of a manuscript requires an unusual combination of knowledge. Arnold Relman, commenting on one major plagiarism case, points out that a "vast literature" had been involved in the questionable articles; he believes that most reviewers fail to detect plagiarism because they simply do not "carry around a perfect memory of every paragraph that appeared in previous articles."[78] And, indeed, the literature in any one field or subfield of science is huge. In 1980, for example, *Chemical Abstracts* monitored 12,728 science and technology journals relevant to chemistry research and they published 475,739 abstracts on chemistry topics alone (73% of which were abstracts from the journal literature).[79] Even a small subspecialty area from that data base each year contains thousands of abstracts.

The more referees there are, of course, the more likely it is that one of them will detect a problem. Nevertheless, referees cannot be *expected* to recognize deception. When detections are made, they may seem serendipitous discoveries, but many analysts argue that serendipity plays little role and that referees usually detect fraud only "on the basis of private knowledge." No data exist to support either conclusion.[80] Once forgery or plagiarism has been caught, it is easy to conclude in retrospect that referees should have seen certain gross implausibilities or inconsistencies in the data. In one sensational case, a well-known editor has stated that the paper in question described results so "dramatic" and "so quickly obtained" that everyone should have been suspicious (but, in fact, no one voiced suspicions publicly at the time).[81]

Even a skeptical reader may initially accept forged data for all sorts of understandable reasons. Referees who endorse a manuscript's conclusions may subconsciously want the evidence to be correct; the conclusions may fit common cultural biases and the "facts" may be cleverly constructed to those expectations. Editorial practices may themselves hinder discovery of fraud, as in the reluctance to encourage aggressive reviewing. Under most circumstances, reviewers will not see the raw data or artifacts so they must rely on the authors' accounts; when those accounts are skillfully constructed lies, referees and editors alike may be duped.[82]

In some cases, referees, in retrospect, have appeared not just insufficiently skeptical but downright lax. Those who have reanalyzed Cyril Burt's work point out a number of obvious flaws; "many of the aspects of the data are too statistically good to be true."[83] Scientists who at first made "charitable interpretations" of Burt's work later complained that "even the most statistically stupid undergraduate" could have done "a neater job of faking his quantitative results."[84]

Cyril Burt was a British psychologist who, from the 1910s until well into the 1960s, supposedly carried out research in various institutions on the inheritability of intelligence, using sets of identical twins reared apart. Few questions have been raised about Burt's research before 1943, but several social scientists reviewed his later work and have argued that Burt often wrote about experimental subjects who never existed. His deception was so skillful, his explanations so plausible, and his conclusions so politically welcome to certain audiences that Burt's work received inadequate criticism when it was first published. Detecting inconsistencies in his work was easier when scholars could examine the entire body of work at once in the 1970s.[85]

Burt also attempted to cover his tracks. When colleagues requested copies of data that he had described but which did not exist, Burt apparently "reconstructed" it from the published correlations, essentially "working back from his answer to create wholly fallacious data" that were so plausible that they were frequently reproduced in major psychology texts.[86] He also made discovery more difficult by appearing to have relied on secondary data published in uncataloged theses or dissertations, or in obscure government reports; many of the sources cannot now be located.[87] Some fabricated data, supposedly developed in follow-up studies, were also apparently extrapolated from real data collected in the 1910s and 1920s. One historian concludes that, even given the most charitable interpretations of the facts, Burt appears to have known exactly what he was doing. His manipulation succeeded primarily because he was so skillful in the pretense and in a prime position to defend against discovery.[88]

Reviewers may be especially reluctant to question a manuscript when they know the author is a well respected and powerful scientist, or if the work intentionally synthesizes others' work. One journalist writing about the Shervert Frazier case explains that the psychiatrist's plagiarism went undetected for so long because (1) he "borrowed" from researchers whose work was not well known; (2) he published in journals that "were not particularly high-ranking or widely distributed" and he wrote "review papers that did not contain original research"; and (3) the articles in question had also been invited submissions and therefore "were not routinely reviewed by other experts to the degree that papers submitted to [some other journals] would be."[89]

As great a long-term danger, however, lurks in the other direction. Arnold Relman observes that:

If you take an adversarial approach to the review of manuscripts, the system would come to a halt. That doesn't mean to be naive, but once suspicions are aroused, what can you do? You can't hold an inquest. You must discreetly ask for more information. And if the author produces it, you have to accept it. The whole system is based on trust. You can't run science the way you run a criminal investigation.[90]

What obligations, then, do referees have? One suggestion is that referees should be encouraged to "bring prior publications of the same material to the attention of editors."[91] But do they also have a moral or legal obligation to notify editors of their suspicions?

If they do, what happens once suspicions are voiced?[92] If an untenured or postdoctoral fellow detects fraud, as will be discussed in the next chapter, he or she may be in an exceedingly delicate and potentially dangerous situation.[93] By bringing the problem to the attention of superiors, he or she may be labeled a "troublemaker." Some have even suggested that, in such a case, a referee would be well advised simply to reject a manuscript on other grounds and remain quiet.[94] An alternative action, in such circumstances, might be for the young scientist to voice suspicions privately but informally to the journal's editor and urge another round of intensive reviewing.

What obligations do journals have to their referees? Is this a confidential, protected relationship, according to both moral and legal criteria? Should journals begin to adopt formal agreements (e.g., legally binding contracts that spell out the obligations of both parties and any limitations on protection)? At present, formal legal agreements between editors and referees are rare. If such contracts were routinely imposed, how might they affect the refereeing process and the trust that mediates relationships between editors and reviewers? These and similar circumstances warrant further attention.

REFEREE MISCONDUCT

Referees can also be the villains. In order to conduct their analyses, referees are entrusted with unpublished manuscripts. They are allowed to examine them behind a veil of confidentiality, and they usually prove worthy of trust. Nevertheless, suspicions and rumors persist that referees have stolen from manuscripts under review.[95] Michael Gordon writes that a common concern of authors is "that if referees' identities are not revealed, they can give deliberately biased reports, delay papers and even plagiarize findings and ideas, all in the interests of advancing their own careers at the expense of an author's."[96] Such allegations frequently surface in highly competitive fields or where the financial value of an unpublished manuscript may appear to be related to assignment of a patent.[97] Even without independently verifiable evidence of such misconduct, the persistence of rumors indicates serious concern.

The conviction that this problem is real has led some scientists to advocate open refereeing, arguing that it would increase author responsiveness, prompt rejection of "marginal" papers, and give referees "credit and responsibility for their efforts."[98] Making referees *publicly* rather than privately accountable for their decisions might also help to keep them honest, these critics add.

The procedure of anonymous or blind reviewing which is adopted by many journals is actually relatively new to science. Anonymous reviews of manuscripts, either by concealing authors' names from referees, or concealing referees' names from authors, or both (i.e., "double-blind" reviewing), became accepted practice in science journals sometime after the 1940s, but adoption was not uniform. A 1967 study of fifty journals in seven different disciplines (including the natural and physical sciences as well as social sciences) found that only nine journals (eight of those in sociology) at the time required that an author's name and institutional affiliation be concealed from referees.[99]

The proponents for open refereeing mount strong arguments for their position. "If science is really supposed to be the most honest game in town," one chemist has argued, "then why not referee our publications above board."[100] Many journals publish an annual list of reviewers but names are not associated with particular articles; other journals do publish referees' names along with a published article. One way or another, acknowledging referees gives public credit, assigns responsibility, and increases pressure for constructive rather than destructive refereeing. Most of the fears of anonymous reviewing surround negative rather than positive assessments, however. What might be the consequences of informing the authors of *rejected* articles who the referees were; would that improve fairness and objectivity of review? Reviewers might be so fearful of alienating colleagues that reports would become bland and useless as guides to selection. Similar arguments have been made against open employment and admissions recommendations.

Proposed solutions are as diverse as the criticisms. Stephen Lock, in a terse analysis of peer reviewing, observes that because a "blinded" review process is "impossible to achieve all the time," some people sincerely believe the system is inequitable; but Lock and most medical editors have not abandoned the practice because they maintain that it offers "a partial remedy" against bias.[101] Daryl Chubin and Edward Hackett, in their study of peer review, imply that blind review does just the opposite and is used to obscure bias or to veil inequitable decisions.[102] In my experience as an editor, the risks of error are generally outweighed by the known benefits. On many occasions, I am convinced, double-blind reviewing helped to keep the playing field level for women, for young or unknown scholars, and for those from lesser-ranked universities.[103] Longitudinal studies that compare acceptance rates by gender for comparable groups of peer-reviewed journals would help to test the effectiveness of "blinded" versus "open" review in fostering equity.

Despite the fears voiced about the trustworthiness of referees, few formal allegations of theft by referees have actually surfaced through the years.[104] Because of the protective layer of confidentiality that surrounds refereeing, an author may discover a theft only after the plagiarist has published the stolen text or ideas. The difficulty of proving allegations of referee misconduct was emphasized in one controversial case that involved apparent theft of ideas, research methods, and experimental approaches during the refereeing process.[105] The dispute began in July 1986, when Robert R. Rando and two other colleagues at Harvard University submitted a manuscript to the *Proceedings of the National Academy of Sciences* (*PNAS*) through John Dowling, also at Harvard, who is a member of the National Academy of Sciences and a *PNAS* editor.[106] Dowling sent it for review to C. David Bridges, a biologist then on the faculty of Baylor College of Medicine in Houston and also a specialist on the topic. Bridges returned the manuscript a month or so later, saying that he could not review it because he was working on a similar experiment. The next spring, Rando (whose article had since been accepted and was scheduled for an April issue of *PNAS*) noticed an abstract of a talk to be presented at a biology meeting. The abstract described work that appeared to be "strikingly similar" to that of the Harvard team and so Rando contacted Dowling, who said that the abstract's author (Bridges) had indeed been a referee on the original manuscript.

The dispute at first seemed to involve uncomfortable coincidence and simultaneity. While the Rando team had been readying their article for *PNAS*, Bridges had been engaged in his own effort to publish. He had submitted an article on his research to *Nature* in November 1986 but the article had been rejected; Bridges had then sent a revised version to *Science*, which accepted the article and scheduled publication for June 1987. In the meantime, Bridges had prepared the abstract for an upcoming scientific meeting. The Harvard group was mildly concerned when they learned about the abstract, but news of the forthcoming *Science* article spurred them into action, so they complained in writing to both *Science* and *PNAS*.

Behind the scenes, in a meeting brokered by Dowling, Bridges met with the Harvard researchers and reportedly agreed that his article would reference the Harvard work and would acknowledge its priority. Up to this point, all allegations had been raised privately, within the confines of the publishing arena, and discussions involved only editors and authors. According to all accounts, no one had yet formally alleged plagiarism or theft of ideas or other serious misconduct.

When Bridges's article appeared in the 26 June 1987 issue of *Science*, it referenced the Harvard work only in a footnote. By that time, the gossip network in the scientific community was also buzzing. Rando and his colleagues next contacted Bridges's employer, Baylor University, which then notified NIH of the accusation (because Bridges's work had been funded by the National Eye Institute). The resulting government investigation was completed in June 1989, when an NIH Special Review Panel issued a report unfavorable to Bridges.[107]

Although Bridges has strongly objected to both the NIH investigation and its conclusions, the report of the NIH panel was unambiguous in condemning his conduct. The panel found Bridges guilty of "patterning his experiments after those detected in a manuscript sent to him for review, using the information to publish his own work and falsifying his records to claim credit" for a discovery.[108] The NIH report referred to "wanton abuse of the peer review system with the clear intention of gaining personal aggrandizement."[109] As punishment, the NIH proposed to bar Bridges from serving as a proposal peer reviewer for ten years, stripped him of current grants, and recommended that the Secretary of the U.S. Department of Health and Human Services (HHS) permanently deny agency funds to Bridges. It should be noted that Bridges maintained his innocence, rejecting the conclusion as "outrageous" and reached without due process.[110] He appealed the proposed debarment in an HHS administrative hearing (June 1990) but the ruling was not reversed.

The NIH and university investigations represent one part of this controversy; the circumstances surrounding the journal publication process represent another. *Science*, in particular, found itself in the middle of an embarrassing situation. After release of the NIH decision, the journal issued a carefully worded news release.[111] *Science*'s editor, Daniel Koshland, also drew criticism, including pointed commentary from a member of Congress, for his failure to address the original 1987 allegations.[112] Koshland justified his publication of the Bridges paper, even in the face of questions raised before publication, on the grounds of "the very exciting results" it described. He argued that the dispute had at first seemed to be over "priority," not plagiarism. It seemed "more harmful to hold up the paper than to publish it," he explained.[113] As the journal's news release stated:

> We and our reviewers recognized the paper . . . as a preliminary report with a potentially exciting conclusion. Shortly before we published it, we became aware of the controversy . . . , but we decided that we did not have sufficient evidence to take action with regard to the controversy.[114]

In a later editorial, Koshland steadfastly defended his handling of the original accusations. At *Science*, he wrote, standard practice is that "if a paper published elsewhere reporting the same results appears while our manuscript is under review, that published paper cannot be a factor in . . . decision-making." *Science* prefers to "treat the two papers as independent discoveries of the same finding." Because the questions raised during the *Science* review process involved "proper credit," not plagiarism, the journal proceeded as usual. *Science*, Koshland reaffirmed, would "continue to expose misconduct when it occurs," but it would also "assume good faith unless presented with evidence to the contrary."[115]

In discussing this particular case, many journal editors admit that such disputes are not rare, especially when issues of plagiarism and priority are entwined. The informal adjudication that Dowling attempted in the beginning of the dispute also reflects the normal course of events. Not all editors agree, however, with the course of action chosen by *Science*. There is considerable criticism, most surrounding the argument that, once informed of allegations, *Science* should have delayed publication.

This case has helped to draw attention to the possibility of misconduct by manuscript reviewers and to the lack of uniform guidelines for refereeing. It might also stimulate the development of more explicit policies for handling allegations of this type of misconduct. Journals should, at the least, begin to consider how allegations might be handled and to discuss referees' responsibilities with them beforehand. It also emphasizes the need for scientific associations and publishers to pay more formal attention to developing equitable standards and guidelines for refereeing.[116]

EDITORIAL MISCONDUCT: *LUCKY JIM*

If theft of ideas by a referee is reprehensible, plagiarism or fraud by an editor is even worse. To date, despite many private allegations, few cases of misconduct by editors have surfaced publicly, other than that of Cyril Burt (who allegedly abused his position as editor of a psychology journal by ignoring "normal processes of peer review" and publishing, under assumed names, material that commented on his own research).[117]

The absence of well-documented cases of editorial misconduct could indicate the absence of problems, but more likely it reflects the fact that aggrieved authors probably feel powerless and so fail to report their problems with editors. To whom does an author complain? One particularly lucid account of such a dilemma occurs in a popular British novel from the 1950s. As Agatha Christie reminds us: "Fiction is founded on fact . . . But is rather superior to it."[118]

Kingsley Amis's *Lucky Jim* revolves around the rebellion, hopes, and jaded optimism of an unlikely hero, Jim Dixon, an untenured lecturer in history at a small English university.[119] Dixon's attempt to publish his manuscript, "The Economic Influence of the Developments in Shipbuilding Techniques, 1450 to 1485," forms a crucial subplot to the novel. In a forerunner of "publish-or-perish" competitions, Dixon has been advised that he must publish something to have any chance of being retained at the university. If a journal (*any* journal) accepts Dixon's article, then his mentor can use publication as a yardstick of the young historian's legitimacy and promise—or so Dixon is led to believe.

In describing Dixon's work, Amis performs a masterful hatchet job on writing in the social sciences, for the young historian has constructed an article that even *he* calls full of "frenzied fact-grubbing and fanatical boredom." Its title, Dixon reflects, "crystallized the article's niggling mindlessness"; its "funeral parade of yawn-enforcing facts" threw "pseudo-light . . . upon nonproblems."[120] Dixon regards himself as a "hypocrite" for having produced it. It is, predictably, rejected by one journal; so Dixon desperately tries another, a new journal just advertised in *The Times Literary Supplement*.

Like many academics, Dixon can only contemplate publishing from the outside: the process is a black box, and editorial review, a game. He calculates that because a new periodical will not have a long queue of manuscripts, his chances of acceptance may be better; but when weeks pass by without a response, his optimism twists into anxiety.[121] Then Dixon's mentor repeats some gossip he has heard about the editor—there was "something shady" in his past involving a "forged testimonial," he murmurs. Dixon panics and calls the editor, coyly saying he was just "wondering what was going to happen to that article of mine you were good enough to accept for your journal."[122]

The editor hedges. Things are "difficult," he explains. There is a long queue of material. Typesetting is a very complicated procedure. It's an exquisite parody of editorial excuses. The editor berates Dixon for his questions, and for bothering him: "I know chaps in your position think an editor's job's all beer and skittles; it's far from being that, believe me."[123]

When Dixon presses, he responds: "I can't start making promises to have your article out next week . . . You don't seem to realize the amount of planning that goes into each number, especially a first number. It's like drawing up a railway timetable."[124] The editor rings off.

Dixon's anxiety turns to helpless anger a few weeks later when he sees an historical survey of shipbuilding techniques in an Italian journal.[125] Even without fluency in Italian, Dixon can tell that the article is a close paraphrase, possibly a verbatim translation, of his own. The author is the editor of the new journal. Dixon laughs hysterically—intellectual treachery! That agonized bleat of betrayal signals a breaking of trust undeserved by even the unlucky Jim Dixon.

RECIPROCAL OBLIGATIONS

Should editors assume the role of moral police or "ethics watchdogs" over the research discussed in the pages of their journals? During the time when regulations governing the use of human subjects in scientific experiments were being developed in the United States, just such a debate took place over the appropriate role for editors in monitoring adherence to the spirit as well as the letter of the law. Those discussions provide some useful comparisons for the debate today over editorial roles in preventing or detecting data fabrication, forgery, or similar unethical acts.

In the 1970s, there was concern that although not every human subject at every institution would be covered by the proposed regulatory scheme, every experimenter should conform to the spirit of the regulations. The suggestion was made that editors of biomedical journals be enlisted in "prepublication ethical screening." Two proponents of this approach, Yvonne Brackbill and André E. Hellegers, argued that editors should willingly shoulder some professional "responsibility for ethical monitoring" even though the research institutions had the ultimate legal responsibility for surveillance.[126] To get some indication of the extent of support for the idea, Brackbill and Hellegers conducted a survey of editors of medical journals.[127]

They found that a majority of editors believed that authors should be required to submit certification of Institutional Review Board (IRB) approval along with their manuscripts. Many editors also stated that they would refuse to publish articles based on research that violated the various ethics codes. Nevertheless, a significant proportion believed that editors, in judging manuscripts, should not weigh ethical considerations equally with such factors as methodology or style.[128] The discrepancy in opinions was noteworthy because, although these same editors undoubtedly condemned the abuse of experimental subjects, they were unwilling to sanction the use of journals for policing ethical conduct within the laboratories.

Many editors espoused exceedingly conservative approaches, placing strict limits on what should come under an editor's control. In a group of commentaries accompanying the Brackbill and Hellegers report, several editors examined the implications of aggressive ethical monitoring by journals. Arnold S. Relman questioned the inherent practicality of the suggestion—how could any editor know about such behavior, or determine whether an author was telling the truth?[129] What if an author's institution or nation were not governed by formal rules or regulations for treating human subjects, he asked? Social science editor Eleanor Singer wondered whose ethics would be "adopted as a standard" in such cases. There might be consensus at the extreme end of abuse, she pointed out, but there would understandably be less agreement in the gray areas.[130] The norms at issue were inherently in conflict, she noted, because there were three types of decisions at stake to which these ethical criteria might be applied: "(1) the research I personally do; (2) the research I would approve as an IRB member; [and] (3) the research I would publish, as an editor." In the first role, one serves the norms of intellectual, educational, and professional groups; in the second role, decisions must be governed by an understanding of the purposes of federal regulations. In the third role, decisions will depend on how each individual defines the editor's job; for example, to many editors, the work may be "to disseminate verifiable knowledge, obtained by scientifically appropriate and replicable methods," not to act as a police officer.

Journals in the biomedical and social sciences did, in fact, play crucial parts in publicizing the rationales for IRBs and for helping to build a consensus of opinion behind them. And many large biomedical journals actively policed other types of ethical behavior. In 1985, for example, *JAMA* began requiring authors to list "all affiliations with or financial involvement in any organization or entity with a direct financial interest in the subject matter or materials of the research."[131] In announcing this new policy, *JAMA* stressed that information would be held in confidence, would not be revealed to referees, and would be stated in any published article only in a manner agreed upon by the author and the journal. John Maddox has articulated a less rigid position for his journal with regard to an even more controversial issue, the experimental use and treatment of animals: *Nature* does not automatically decline articles from countries known for their less stringent animal care regulations.[132] Maddox has also said that *Nature* would not automatically reject a paper that made unfair and unauthorized use of another's data, although the journal would "reserve the right to point out to readers what had happened, and to allow the aggrieved party to comment."[133]

Research fraud and other ethical misconduct by authors and referees represent types of behaviors in which journal editors simply cannot claim disinterest. As the number of publicized cases of fraud and misconduct have increased, many editors have adopted aggressive approaches and advocated greater editorial responsibility in tracking down the truth; many have argued that journals should take a leadership role in combating fraud.[134] For some, the preferred solution is a "dynamic response" in which the editor first informs the accused of an allegation and then, if no response is forthcoming, contacts the accused's employer.[135] Still others prefer a more relaxed approach, citing Rudolf Virchow's statement, "In my journal, anyone can make a fool of himself."[136] A few editors remain passive, content to let others discuss the issues or to investigate allegations, but they are increasingly in the minority. The lack of agreement on the degree of appropriate responses reinforces the need for open discussion of where the lines of responsibility should be drawn and whether some uniform policy should be imposed on all publishers and editors and, if so, by whom.

Where does the responsibility for thorough evaluation lie? An obvious response is: "everywhere." All parties involved in research communication bear equal responsibility for diligence and care in avoiding error and in acting honestly. The first line of defense must be the offices and the laboratories in which research is conducted. The ad hoc faculty committee at UCSD that investigated Robert Slutsky issued a number of useful recommendations for closer supervision in laboratories, and they urged scientists to take more responsibility for assuring accuracy in their work. Their recommendations for journals, however, seemed based on an unrealistic assessment of culpability—one that perpetuates a troubling myth about the control of content. "Journals should not be so insistent on publishing short papers," they advised, "since this encourages fragmentation and republication of data."[137] Such admonitions attempt to locate blame for wrongdoing outside the laboratory or faculty office, yet *authors* are the ones who fragment and republish. Editors may accept shorter articles, but none encourage multiple publication of the same thing and many journals have policies prohibiting it.

The UCSD committee also urged that journals use "more care" in evaluating statistical data in manuscripts, but offered no ideas for how to structure, much less pay for, additional technical review. Again, the institutions with which researchers are affiliated could offer review, or even require it, before manuscripts are submitted.

It is always easy to judge actions in hindsight. As journalist Barbara

Culliton has emphasized, the appropriate "role of scientific journals in weeding out suspect papers" is to act cautiously.[138] And yet the cases outlined in this chapter and elsewhere in this book show repeatedly how the same situation can draw conflicting reactions. If no formal allegations have been made, if there are insufficient grounds for suspending publication, and no stated policy seems to apply, then the appropriate editorial response may not be obvious.

Whatever the ultimate choice, editors do have a responsibility to protect the innocent and the wrongfully accused. On occasion, that responsibility may require them to proceed with the publication process fairly and efficiently, if cautiously. Suspicions should not be acted upon as if they constitute proof, even though that policy may well result in publication of articles that are later repudiated.

A crucial route to reform clearly lies in revisions of peer review procedures, but those changes must originate within each individual journal and must suit each journal's standards and idiosyncratic review systems. If more editors explicitly state their own standards and expectations and share in the formulation of guidelines, some aspects of the "system" might be reformed from within.[139] Radical changes in how publishing decisions are made is unlikely to *prevent* unethical conduct by individuals, but sensitive reforms will certainly help to alleviate the misunderstandings that frequently hinder prompt detection, investigation, and correction of malfeasance.

Regardless of the specific review system employed, maintaining a reasonable balance between discretionary choice and informed judgment requires that editors first assume that all authors and referees act honestly. The process requires its participants to adopt reasonable skepticism; it relies on relentless scrutiny of ideas, methods, and writing quality; and it demands that neither referee nor editor will abuse an author's trust. As John Maddox observes, "Erring referees should reflect that they are usually authors as well."[140]

6 Exposure
The Whistleblower, the Nemesis, and the Press

It's an unpleasant business. More like some detective's job . . . [A]s historians we've got to tell the truth about the past as far as we know it, but that's quite a different thing from searching into the truths of people's lives here and now. All this prying and poking about into what other people prefer to keep hidden seems to me a very presumptuous and dangerous fashion.

—Angus Wilson[1]

Science cannot tolerate fraud, but it should not be at the mercy of headline-happy journalists or incompetent whistleblowers.

—Daniel Koshland, Jr.[2]

Exposing fraud and deception in professional conduct brings no applause, few rewards, and little public satisfaction. Understandably, even scientists who are morally repulsed by unethical conduct sometimes remain quiet lest they appear "headline-happy" or fanatical. Yet, time and again, a few individuals do pluck away at indifference, prod sluggish bureaucracies, and publicize what they sincerely believe is unacceptable wrongdoing. They appropriate this task with no official status (and, in many cases, with disapproval); sometimes, they conduct an informal investigation single-handedly, with no support from colleagues, administrators, or officials. As their efforts continue, they attract vigorous criticism, especially if their targets are senior scientists. They may even be accused themselves of hidden motives, of secretly harboring revenge. If their allegations later prove to be correct, society will probably applaud their courage and persistence—but it will be a bittersweet victory.

The circumstances of professional engineers or office workers who blow the whistle on unethical or illegal activities in contracting, construction, or industry are well documented; much of the insights derived from that literature are relevant to the issues confronted in the scientific setting.[3] When individuals attempt to publicize fraud or misconduct in scientific publishing, they face a special set of problems, however. If accusations are to be taken seriously, for example, they must be investigated without endangering the confidentiality of various participants in the publishing process. This chapter focuses its attention on the individuals who bring charges, who cause them to be brought, or who publicize them. Who are such individuals likely to be? What spurs them on? What conflicts of interest, pressures, and compromises do they confront?

Traditional *whistleblowers* play crucial roles within the scientific community in uncovering professional misconduct. They are insiders, employed (or recently employed) within the organization or institution under investigation. They may have observed the questionable conduct as a co-worker, or may have noticed inaccuracies in a student's or supervisor's data. Rarely do whistleblowers attempt to conduct more than a perfunctory investigation; instead, they voice their suspicions or pass on information to an official who can investigate or to someone, such as a more powerful colleague, who may pressure for an investigation. On occasion, a whistleblower may go outside the organization and contact journalists or others (e.g., politicians) who can bring publicity and can pressure for change. Their roles are usually quite limited, however, after those in power take the allegations seriously. Even though whistleblowers may have initially drawn the attention of administrators or government officials, they rarely participate in the formal investigation other than as witnesses.

In the 1970s and 1980s, studies of whistleblowers in engineering tended to examine the factors that facilitated or inhibited their activities, and the economic, ethical, or political effects of their actions. Discussion emphasized the divided loyalties they may experience, especially if they are afraid that blowing the whistle may inadvertently harm co-workers or the reputation of an organization in which they believe. In the context of scientific publishing, whistleblowers might be members of an editorial staff who bring allegations about an editor or may be authors who uncover wrongdoing by coauthors. In most cases, these people experience conflicts of loyalty and conscience similar to those experienced by any employee who discovers reprehensible conduct among co-workers or employers.

In many of the well-known scientific misconduct cases, however, a quite different type of person has appeared. These participants—*"nemesis" figures*—come from outside the organization or laboratory involved, although they may have been notified of the conduct by someone within the organization. They act unofficially, with no community support and sometimes with hostile condemnation from colleagues. With little fanfare, and occasionally with obsessive secrecy, they meticulously trace the extent of the misconduct and assemble their evidence. Once convinced that some violation has occurred, the nemesis usually pushes for an official investigation but will readily bring the problem to the attention of the press if an official investigation (or resolution of it) appears stalled. Unlike whistleblowers, the nemesis figures not only have had no prior connection with the accused (although they may share similar scientific training or research interests) but they also rarely display personal animosity toward the accused. Instead, they exhibit rigid integrity, an intense feeling of moral outrage, and they perceive themselves as acting out of social responsibility and commitment to community standards. A nemesis stands in, as moral judge and jury, for the research community as a whole. At the extreme, they appear uncompromising, almost fanatical, and may seem unconcerned about collateral damage to innocent people caught up in a case. At the other extreme, they may be overly sensitive to the reactions of others, almost pathologically hurt by their experiences but nonetheless convinced of the moral correctness of their actions. They believe that exposing wrongdoing is a noble action.

Nemesis figures are most likely to act where no routine procedures exist for bringing formal allegations, or where existing policies fail to encourage investigation. The nemesis appears to be convinced that if he or she does not act, then no one will. And if no one acts, then an obvious violation of professional norms will go uncorrected, uncriticized, and unpunished.

THE NEMESIS IN LITERATURE: *ANGLO-SAXON ATTITUDES*

Novelists, being good cultural barometers, often reveal social shifts and cultural change long before they become fodder for sociologists, historians, or political scientists; fiction readers are assisted in "experiencing" emotions or behaviors that later become textbook truisms. Just such insight to the nemesis figure in science may be found in a number of modern novels. Although imaginative literature at heart, these books show how the motives of the accused and accuser intertwine and why the nemesis experiences conflicts similar to that of his prey; they pose explanations that, for real situations, remain hidden in sealed administrative records or locked in the participants' hearts.

Nemesis characters in fiction tend either to initiate or to reconfigure investigations of academic fraud out of some profound search for justice. The characters come from the same world, socially and intellectually, as the accused researcher; they know the idiosyncrasies of the context in which wrongdoing takes place. Perhaps as a consequence of this empathy, they experience severe conflicts of conscience: the investigators seem compelled to discover the truth, even though revealing the truth may embarrass or harm their college, journal, organization, or friends.

A nemesis figure is essential to the classic mystery novel form in which a professional detective or amateur sleuth compulsively sniffs out clues, seeks the truth, and eventually reveals the guilty person to the world. Although Dorothy Sayers's *Gaudy Night* (1936) is better known as a novel about women's education in England, and for the romantic battles of its protagonists, the writer Harriet Vane and amateur sleuth Lord Peter Wimsey, the book also explores the ethical dilemmas presented by academic fraud and cover-up.[4]

Harriet Vane has been asked by her Oxford alma mater to uncover the identity of a prankster whose antics not only shatter the college's peace and contemplative life but, if the newspapers get hold of the story, also could tarnish its hard-won reputation for serious scholarship. No one is exempt from suspicion at the beginning, so Vane painstakingly eliminates the innocent in order to sift out the guilty. When she complains to Wimsey that she feels "exactly like Judas," spying on her former professors, he proffers no consolation. He has stared at the same reality many times: "Feeling like Judas is part of the job."[5] In a previous case, he recalls, his "meddlin'" had created more victims, as the murderer killed repeatedly to keep from being discovered.

Investigations can harm innocent people; in Wimsey's case they were in essence "sacrificed" for the cause of justice. As Miss Edwards, a biology professor at the college, confirms with biological dogma of the 1930s: "You can't carry through any principle without doing violence to somebody. Either directly or indirectly. Every time you disturb the balance of nature you let in violence."[6]

As the two detectives discuss the case with the college dons, they draw out parallels between their work and scientific research. Both detectives and scientists use facts to establish a conclusion; in the ideal, they do so independently of what will be made of the facts, by either the courts or society.[7] This similarity is one that causes considerable difficulty today when lawyers or government officials enter university investigations of research fraud, because academics are frequently unwilling to consider the plausible outcome: criminal conviction, fine, or imprisonment.

The characters in *Gaudy Night* discuss how investigations inadvertently upset the delicate balance of collegial relations and raise inevitable conflicts of loyalty. Once an investigation is begun, can it be stopped? What if one knows about a falsification of data but knows that revealing the crime will hurt innocent people? What responsibility then? Can information ever be "forgotten" out of pity for the fabricator or the innocents? No, the dons agree. One must press forward because "to suppress a fact is to publish a falsehood." When higher values are at stake, one's own private feelings must be put aside. Could one ever excuse a deliberate falsification of data, for example? The academics agree that one should not and yet they admit that it is frequently done for the sake of argument or ambition. When, at the end of the book they confront the human face of their tormentor, then they, too, experience the harsh reality of "meddlin'."

The conflicts of conscience that Dorothy Sayers explores in just a few scenes, Angus Wilson draws out as the central plot of *Anglo-Saxon Attitudes* (1956), a novel influenced by the Piltdown case itself.[8] Set in London in 1952–1953, the novel revolves around the legitimacy of a 1912 archaeological excavation on the east coast of England (the fictional "Melpham, East Folk") which had uncovered the tomb of an Anglo-Saxon Christian missionary, Bishop Eorpwald. Because Eorpwald's coffin contained a wooden fertility figure, medievalists regard the site as evidence that Celtic and Christian religions had intermingled during that period. Wilson effectively satirizes the controversy then raging over the authenticity of the Piltdown skull by having the fictional discovery at Melpham serve as evidence for the authenticity of subsequent work, just as Piltdown supported many theories in twentieth-century anthropology.[9]

The novel's central character, Gerald Middleton, a sixty-four-year-old retired professor of early medieval history, regards himself as a failure, professionally and personally, despite the obvious respect of his peers, who have nominated him for a prestigious editorship. He believes that he has engaged in self-deception all his life, failing to admit his true feelings in important personal relationships and evading "the truth, past and present."[10]

When Middleton accidentally encounters vague evidence that casts doubt on the authenticity of the Melpham dig and on the veracity of its principal director (and Middleton's former professor), the late Lionel Stokesay, the conflict he feels is intense. Was Melpham a moment of truth "or a moment of untruth that looks like truth," Middleton muses. The ambiguity justifies his hesitancy; he cannot help but consider the consequences of an investigation. If others were in his place, they would eagerly pursue "a wild will-o'-the-wisp" and would think little of "upsetting the balance of the English historical profession, destroying the reputation of a remarkable historian like Stokesay and God knows what else on the evidence of a few words"; but to Middleton, only a serious infraction of moral norms might justify the upheaval provoked by public inquiry.[11] This is a private matter, a squabble among historians, not a matter for the newspapers or the British public:

> Was it not after all a small point of historical truth that mattered really very little to medieval specialists and nothing, but absolutely nothing at all, to anyone else? So much the more reason for taking no action on so little evidence.[12]

Identifying the likely hoaxer becomes the crucial first step, and here Middleton does have vague suspicions. He remembers meeting with his college friend, Gilbert Stokesay, son of Professor Stokesay, just as World War II was beginning. Gilbert, a poet and essayist, part of the interwar avant-garde literary set surrounding D. H. Lawrence and Wyndham Lewis, was drunk and flushed with fear of going to the front. Gilbert referred to the war as "a mammoth practical joke," the "biggest practical joke of all time," and then boasted that he once accomplished his own "mammoth" hoax. He had actually found the famous Melpham pagan figure at a nearby dig and, "with a little help," secretly placed it in Eorpwald's tomb just before the public opening. He fooled everyone, Gilbert crows, the historians *and* his own father, and he confided that he would reveal the fraud only once his father was knighted.

Gilbert died in the war, so Middleton had always dismissed the drunken "confession" as a joke. Now the story seems plausible. And, if it is the truth, then there is no question about the figure's authenticity, only about when it was placed alongside the Bishop's body.

A "desperate brooding" overcomes Middleton as he stands at the crossroad that all nemesis figures face. The coincidence of the discoveries at Melpham and Heligoland are increasingly cited as proof of the authenticity of both digs: "Two or even one and a half have always a disproportionately greater power of suggestion than one," he observes. The presence of one lie threatens the integrity of an entire body of scholarship, yet publicity will bring other damage.[13]

He turns to the president of the history association, Sir Edgar Iffley, for advice. When Middleton explains the possible role of Gilbert Stokesay, Sir Edgar calls Gilbert a "lunatic" (insanity providing the convenient explanation). Rather than recoiling "with disgust from anything smelling of scandal," as Middleton has assumed he will, Sir Edgar urges action, however, even criticizing Middleton for having delayed the inquiry.

To Middleton, proving the existence of a hoax also means delving into a "far larger question of historical truth." The lie about Melpham, he reflects, "if lie it is, has become the cornerstone on which a whole false edifice [has been] erected. And even if it wasn't so, even if it was just one single historical oddity," then it still should not be treated lightly. "If an historian has any function at all," he argues, "it is to maintain honesty. The study of history can't be the plaything of . . . egoistic mockery."[14]

Even in the hoax inquiry itself, the scholar must maintain academic standards. Without evidence, Middleton can do nothing. So he interviews the friends and relatives of the men involved, seeks out scraps of conversation, memories of intrigue, and eventually locates incriminating correspondence between Lionel Stokesay and his close friend Canon Portway, both now dead. After Gilbert's death, Stokesay had indeed uncovered the truth and shared it with Portway; but, ostensibly to protect the prestige of their local amateur history group, they had destroyed the evidence and agreed to mutual silence.

Notified that there is conclusive proof of a hoax, Sir Edgar appoints a special committee to handle the public announcement. Once revealed, the truth must be accepted, the scholars conclude; but it need not be waved about. At all costs, they must control the newspaper coverage "to protect scholarship from ridicule." Such information cannot, of course, really be kept quiet. The community of trust is disrupted, the conspiracy of ignorance is shattered, and the nemesis finds little praise from his colleagues for his truth-seeking.

REAL LIFE EXAMPLES

Whistleblowers and nemesis figures also populate many of the actual cases of plagiarism and fabrication, sometimes playing an overt, aggressive role (including interactions with the press), other times remaining hidden in the background.[15] If a nemesis figure has had no previous relationship with the accused and is not connected to the accused's institution, then a course of action (e.g., who should be notified first, and how) may not be obvious. When Linda Ferguson and Janet Near analyzed young whistleblowers in science, they found that young scientists appear especially ill-equipped.[16] In contrast to whistleblowers in other fields, the scientists often did not know "what channels to use" for making allegations and did not feel confident that their accusations would be treated fairly or seriously. Other work by Near and her colleagues shows that, in general, people in all professions are more likely both to "observe and [to] report wrongdoing if they believe that whistle-blowing is a positive force."[17] Those who do blow the whistle tend to perceive "effective change" as their goal and will sometimes wait before reporting their suspicions in order to avoid retaliation.[18]

Not all young academics are timid, of course. In 1988, when Paul Scatena, a graduate student in philosophy at the University of Rochester, began to question the authenticity of the publications of a prominent Harvard professor of psychiatry, Shervert Frazier, he chose to exploit several different routes for exposing the truth.[19] Scatena had been reading psychology journals related to his thesis topic when he noticed similarities between review articles by Frazier (some dating back to the 1960s) and papers written by other psychiatrists. Instead of just filing his suspicions away, Scatena began to build a "dossier" of proofs, carefully tracking down "unattributed sources for every paragraph" of three of Frazier's articles.[20]

Once that evidence was assembled, Scatena decided to take his concerns first to the scientific community, so he wrote to *Science*, which he considered a "bulletin board" for the scientific community. He also wrote to Frazier's employer, Harvard Medical School, mentioning that he had notified *Science*—"I was afraid if I just sent it to the medical school they might ignore it," he later explained; "I wanted to make sure I wasn't ignored."[21] If Harvard had not acted on his information, he stated later, then he would have taken his claims to the mass media.[22]

When reporters asked why he went to so much trouble, Scatena replied that it was not "personal": "I don't know the man. But think of it—a major figure in psychiatry who was plagiarizing."[23] In fact, this was not the first time Scatena had taken on a nemesis role. He told *The Boston Globe* that "he had previously found plagiarism in a paper submitted by a philosophy graduate student at another university and turned in the wrongdoer, forcing retraction of the paper."[24]

Scatena's willingness to be interviewed and to "go public" with his allegations was somewhat unusual. More often, the nemesis acts behind the scenes, making telephone calls or writing letters in an effort to initiate or urge an official investigation but not necessarily seeking publicity. In this respect, nemesis figures appear to share the goal of protecting the scientific community's public image. They are part of the same world.

On occasion, a senior scientist or scholar may even act as a type of "stand-in nemesis," communicating allegations to administrators or government officials which have actually been made by a young (or otherwise less powerful and more vulnerable) student or colleague. The original accuser may have observed a wrongful act but is afraid of recrimination. In such a case, the nemesis simply picks up a moral baton and finishes the race. At Stanford University in the early 1970s, for example, a postdoctoral fellow was convinced that his former supervisor had never performed certain key experiments but was afraid to blow the whistle; his new faculty supervisor took up the case and ensured that university investigations were continued.[25]

In the case involving accusations against paleontologist Vishwa Jit Gupta (chapter 4), his former coauthors and collaborators have publicized their allegations, apparently out of chagrin for having failed to question the work earlier.[26] John A. Talent explains: "Few scientists have the inclination to 'blow the whistle' on fellow scientists' obviously spurious data, or have the necessary expertise with which to indulge in psychiatric explanations of research behavior. They choose to ignore such problems."[27] Talent adds that the coauthors' initial reluctance "to go public . . . allowed the edifice of disinformation to reach monstrous proportions"; but he rejects the criticism that whistleblowers tend to exaggerate potential problems. He and his colleagues, he asserts, "deliberately *underplayed* the situation" by initially displaying only "a sample" of the evidence they had. Now they believe the extent of questionable evidence to be far greater than originally suspected. Does such a shift—from uncertainty to conviction—represent a typical progression of attitudes in these cases? It is possible that nemesis figures feel guilty for suspecting colleagues they once trusted and so they attempt to absolve the guilt by locating far more evidence than necessary to prosecute or convict. It is also possible that publicity about a case simply begins to attract evidence from other sources.

PUTTING PUBLIC FACES ON PRIVATE MOTIVATIONS

It is not unusual for an individual to adopt the nemesis role repeatedly. One of the most controversial figures of this type, Walter W. Stewart, first became known for his attempt to publish an analysis of John Darsee's coauthors; a few years later, he participated in another sensational episode, the *Nature* investigation of Jacques Benveniste.[28] In the course of the Benveniste inquiry (see chapter 7), Stewart's involvement in an even earlier *Nature* investigation of fraud became known.

Walter Stewart and his research collaborator Ned Feder have now become household names within the scientific community. Analysis of the news coverage of the pair over a five-year period shows that, in fact, perceptions of them were not static—the more active they became, the less favorable the coverage. In 1985–1986, as they were attempting to get their "Darsee coauthors" manuscript published, descriptions of them were straightforward and slightly sympathetic. They were portrayed as "career scientists," "determined" to get published, sincere if naive chaps who were fighting a powerful establishment. In Spring 1988, after the *Nature* article was published, they asked the NIH to detail them to a congressional oversight committee, where they began to work on investigations of other scientific fraud cases. The tone of news descriptions of them changed. Some accounts implied that the pair gained personal gratification from their activities on Capitol Hill. The *New York Times* called them "critics of science," who were "performing a valuable service in rooting out sloppy and even deceptive practices," but added that they were "having the time of their lives as they track down errors." Stewart was quoted as calling their experiences "absolutely exhilarating" and "extraordinarily exciting": "It's really a lot of fun. We wake up every single morning and wonder what's going to happen today."[29] The article described Feder as "scholarly, intelligent, reticent" and Stewart as having the "air of a brash, bright graduate student tweaking the nose of the scientific establishment."

Editorials and news reports in *Science* and similar publications began to refer to them as "whistleblowers" and, indirectly, to link their investigations to "McCarthyism"—an inflammatory characterization that itself drew criticism.[30] Throughout 1988, journalists called them "critics" (either "outspoken" or "regular" critics); and the adjectives "self-appointed" and "unofficial" appeared. Even the normally sympathetic Daniel S. Greenberg observed that they had "doggedly . . . set themselves up as the unofficial ombudsmen" for science.[31]

In 1988, *Science's* news columns began to repeat Washington gossip that "around NIH, their recognition on the Hill as self-appointed guardians of scientific purity is referred to as the 'lionization of the turkeys.'"[32] A single article contained the statements that they had "assigned themselves" their roles, "entirely on their own initiative," were "self-appointed guardians of scientific accuracy," with "no official standing," and had a "sense of persecution."[33] Then, in late 1988 and 1989, as the cases that Stewart and Feder had been investigating unraveled, and many of their charges appeared to be correct, the hostile language softened—the team suddenly became refered to as "gadflies," seen again as agents of positive change ("fraud busters" and "science's policemen").[34]

At no time have Stewart and Feder ever been portrayed as having powerful positions; more often, they are characterized as slightly outside the ranks, not quite disbarred but disreputable, and as possibly failing to meet their intellectual obligations as researchers because they are preoccupied with other activities. In 1991, a *Washington Post* profile cautiously praised their motives but described them as marginal to science ("iconoclastic, socially reclusive," untypical, "affable" but "disheveled" and out of step with fashion, both social and scientific). Stewart's "flair for scientific detective work" was explained as "driven by a rebellious disregard for things like approbation and status."[35]

What motivates a nemesis to risk professional censure? First, nemesis figures appear to regard themselves as part of the same world as the violators they condemn; they may even identify with them, in the psychological sense, especially if they are members of the same research field. Some nemesis figures have been of comparable or higher status in a field than the alleged violator; but even if they are less powerful, they do not seem to acknowledge the difference in status as relevant. Because ethical codes and professional standards express the mutual trust of all colleagues, a nemesis believes that, by breaking the code, the accused has also broken the trust agreement implicit among all who espouse the code. It is a curiously egalitarian theme. The nemesis sometimes appears to be disgusted with the accused's failure to meet these expectations and, in this attitude, shares Poe's "horror" and offense at the "debasing propensity to pilfer," much as a *grande dame* huffily dismisses a social parasite.[36] As has been said about hoaxers in general, "In the end the striking thing about the hoaxer and the hoaxed is not the gulf which separates them, but the similarities between them."[37]

Second, the nemesis simply *cannot* stay quiet. Once truth is known, it must be revealed. The personality type may represent the opposite of what psychologists call the "bystanders," those people who watch a crime yet do or say nothing. As Leon Sheleff describes in his work on the bystander phenomenon, the motives that compel someone to act, rather than to stand by, are "far more complex than would appear on the surface" and involve "a mixture of self-seeking and self-sacrificing."[38] Even if personal inconvenience, physical danger, or risk to career appear likely consequences of involvement, the nemesis cannot remain still.

Third, unlike some social actors, the nemesis may also feel more vindictiveness toward the transgressor and all he represents than empathy for the party or parties wronged. When nemesis figures have been engaged in attacking plagiarists, for example, they have not demonstrated any unusual concern for the original authors of the text stolen or for the general subject of intellectual property rights. John Maddox states that even though he regards himself as a "supporter" of Stewart and Feder, he finds them "a bit insensitive of other people's feelings. They make no allowance for human frailties like pride, vanity, even vainglory. They are people who believe that everything ought to be a matter of fact. And the real world, even the scientific world, isn't like that."[39]

And, fourth, nemesis figures may also regard themselves as self-sacrificing representatives of the group ("someone must do it," the person thinks), especially if no precedents for investigation have been set, if no formal procedures exist for structuring an investigation, or if official investigations appear to be stalled. A belief that existing rules are faulty or that investigating officials are corrupt may also allow the nemesis to justify unorthodox means of resolution, such as rumor or gossip.

If scientists want to know the truth, then why do they so often criticize the truth-seeking nemesis? In explaining why it took so long to uncover Cyril Burt's fraud, one writer emphasizes that scientists tend to "accept facts as facts" and to challenge the interpretation of facts "rather than their existence."[40] When those interpretations are widely shared, it is hard to challenge them. It is far easier to be skeptical of interpretations that contradict accepted wisdom or consensus. As other analysts explain: "Many scientists want passionately to know the truth. It is only the literary conventions of scientific reporting that compel scientists to feign detachment."[41] The emotional detachment may be far greater than conventionally assumed. The nemesis disrupts the tidy order of science and must be removed either temporarily or permanently from the scientific mainstream. Even though fellow scientists may share the nemesis's conviction that the conduct is wrong, the inquiry itself violates another unspoken norm by disturbing the tranquility and equilibrium necessary to sustain scholarly life.

RISKS AND CONSEQUENCES

A congressional staff member who has investigated scientific fraud has stated that for "every one of the cases we looked at, there was hardly a single person you could point to that ever raised a question and survived."[42] The career of Margot O'Toole, an NIH postdoctoral fellow who became involved in the "Baltimore case," demonstrates dramatically what can happen to a scientist who sees a wrong and cannot let it go. In 1986, while working in the MIT laboratory of Thereza Imanishi-Kari, O'Toole challenged the accuracy of data and experimental results in a *Cell* article coauthored by Imanishi-Kari, Nobel laureate David Baltimore, and David Weaver, another young MIT scientist. O'Toole was eventually vindicated when, in 1991, the results of continuing NIH investigations became public and it was clear that, whatever the truth about the *Cell* article, there had been sufficient cause for investigation after all.

In the intervening years, however, life for O'Toole was difficult. From the beginning of the national publicity to the allegations—following her congressional testimony in April 1988—O'Toole's career was characterized, by supporters and detractors alike, as "ruined." Without any job security, as an untenured postdoctoral fellow, she had questioned the veracity of prestigious and powerful scientists; Baltimore reportedly told her that "he personally would oppose any effort [she] made to get the paper corrected."[43] Her fellowship was not renewed and, for many years, she could not find work as a scientist. It was perhaps easy, therefore, to portray her as vulnerable, "railroaded out of a scientific career," and "obscure," in contrast to her visibly successful critics.[44] Imanishi-Kari publicly questioned O'Toole's scientific competence; Baltimore was quoted as calling her "discontented"; news reports in *Science* implied that, as "a Dingell protégée and whistleblower," O'Toole was stringing out the controversy for her own vindictive satisfaction.[45] The turning point came in March 1991, when it seemed that her suspicions had been correct all along. The *Boston Globe* and the *Washington Post* praised her in editorials; the *New York*

Times called her a "hero," lauding her personal courage and strength of character.[46]

The case of Margot O'Toole demonstrates how ambiguity in institutional policy and procedures, when it hinders an investigation of allegations, can have significant and negative impact on a whistleblower's future employment and professional standing. Within science, the stubborn unwillingness to admit, even at the laboratory level, that the standard checks have failed or that a respected scientist could have feet of clay can stimulate an ironic reversal of the standard response—initial condemnation of the accuser and spirited defense of the accused. Especially within universities, where status hierarchies are rigid and where students and post-docs serve as apprentices, the absence of procedures that are *automatically* triggered by an accusation just raises the stakes and increases the risks. Every senior official who interrogates the trembling accuser becomes another discouragement, another hurdle to the truth. Likewise, disgruntled employees with revenge in mind and no procedures to control them may opt for gossip and rumor—something even the most powerful scientist may find difficult to combat. If formal procedures do not exist, it is harder to force any accuser to place allegations on the record, whether they are truthful or malicious. The situation then works to the disadvantage of everyone, the innocent whistleblower and the innocent accused. As the American Historical Association observes, within universities, there is "a strong tendency to hush up lurid charges, and to rest content with a quiet, equivocal apology because of sympathy for a popular colleague assailed by a remote and seemingly mean-spirited rival."[47]

Community sentiment tends to regard whistleblowers "as more deviant than the people they blow the whistle on."[48] To weather that criticism, they must adopt a pragmatic attitude from the beginning. When he levied his accusations against Shervert Frazier, Paul Scatena apparently just assumed that his actions would be more likely to harm than to help his career: "People say that I did a good thing, but the big-name places may regard me as a troublemaker. I suspect that any big place would not want me."[49] To his credit, Scatena has never indicated that these considerations influenced his choice. He made his charges and stood by them. Margot O'Toole, too, apparently had no illusions about the negative effect of continued protest, despite her defense of the highest scientific ideals in her congressional testimony.

Other scientists who have blown the whistle on supervisors or students demonstrate similar pragmatic assessment of how colleagues will regard their actions; but not all endorse whistleblowing as the best choice. Even seniority provides no protection against the negative backlash against a whistleblower. Although Stephen Breuning eventually pled guilty to federal charges, two of his former collaborators, Robert Sprague of the University of Illinois and Thomas Gualtieri of the University of North Carolina, who had gathered the initial evidence and had brought the case to the attention of the universities and federal agency involved in the federally funded project, were themselves the targets of criticism and retribution as the investigation dragged on for years.[50] Sprague attributes the deferral of his NIMH grant to his role as a nemesis figure; he spoke with considerable bitterness when he testified about the experience in front of Congress in 1989 yet even that testimony drew criticism (and threat of lawsuit) from a university official, a threat recanted sixteen days later with an apology from the university's president.[51]

Jerome G. Jacobstein, whose accusations resulted in a case that stretched from 1981 to 1987, believes that it is extremely difficult "to get a fair investigation of the facts unless there is incontrovertible proof from the outset."[52] Even with a dogged and persistent personality, with copies of the questionable data books, and with the resources to hire an excellent attorney, Jacobstein's pursuit of the truth took years, and the accused scientist continued to serve on an NIH committee even while he was being investigated by NIH.[53]

Bruce Hollis, who blew the whistle on a senior colleague at Case Western Reserve University, suffered even worse treatment. He was fired by the university and criticized in the local news media. Eventually vindicated by an NIH investigation and rehired by the university, Hollis had by then moved elsewhere. "I cannot recommend that junior scientists who discover scientific misconduct blow the whistle," he wrote years later, "unless they want to experience immense personal suffering and a possible end to their scientific careers."[54]

A nemesis figure may have some cause for fearing professional repercussions.[55] Nobel biologist Walter Gilbert, commenting on O'Toole's eventual vindication, warned that "one of the great dangers is that the established community closes ranks against the whistleblower—that's completely what happened in this case."[56] Given the fact that university and federal investigations appear to be taking more time to complete rather than less, a "successful" nemesis must possess extraordinary perseverance, must remain committed to the effort possibly for years, and must ignore sustained personal attacks or acts of revenge, whatever the costs.[57]

The absence of any accepted investigatory procedures for journals sometimes leaves an editor in the nemesis role by default, requiring an editor to judge not just the authenticity and appropriateness of manuscripts but also authors' moral conduct. This circumstance can exacerbate an already difficult situation, for editors caught up in cases of scientific fraud or plagiarism normally experience feelings of betrayal, that is, "the emotional characterization attendant upon discovery that a relationship has not been what it ought, morally, to be."[58]

Should editors accept the nemesis role? Many people argue vehemently that editors should not accept it and that they should, instead, remain aloof and leave the pursuit or investigation to others. In discussing the Bridges case, Daniel Koshland of *Science* writes: "If charges of fraud, backed by documentation, were made against a manuscript under review" in the future, *Science* would not investigate but would bring the allegations "to the attention of those better able to investigate them."[59] Journals, he points out, "have limited investigative capabilities" and "a clear responsibility to preserve confidentiality in the peer-review process."

WATCHDOG OR PIT BULL? THE ROLE OF THE PRESS

Whistleblowers, nemesis figures, and journalists are often lumped together in the same analytical stew, perhaps because the publicity-prone whistle-blowers or determined nemesis figures have so often used the press as the means for drawing attention to their accusations. The role that science journalism has actually played in the fraud controversy, however, has not been either as aggressive as its critics charge or as courageous as the journalists themselves might like to think. They have been drowsy watch-dogs, not hyperactive pit bulls.

Since the 1980s, the news publications that serve the scientific com-munity, such as *Science & Government Report*, *Chemical & Engineering News*, and *The Scientist*, have given more or less consistent attention to scientific fraud and for much of that time their articles have represented the only substantive investigative reporting on breaking stories. Publica-tions like *Chemical & Engineering News* and *Science*, however, are part of the community, and their stories about fraud have been stated in insiders' terms, expressing insiders' concerns. Whenever newspapers like the *New York Times* and the *Washington Post* have picked up the topic, their articles have been phrased not so much as "science" stories as issues of national policy relevance. General newspaper and television reports have tended to place scientific fraud in context, alongside accounts of fraud, waste, and abuse in the aerospace business, the defense industry, and elsewhere. This juxtaposition of professional ethics controversies has helped to reshape the overall tone of coverage, making science seem less "special" and prompting investigative political journalists to look more closely at science as a potential locus of corruption and professional mis-conduct.

The press plays an even more crucial, if secondary role in alerting the political establishment to emerging problems such as that of scientific fraud. As indicators of evolving public concern, press accounts demonstrate to politicians that an issue has taken on wider importance. When it is the actions of government officials themselves that are being reported, the coverage reinforces the impression that the issue is controversial, important, and worthy of public attention. On occasion, a journalist can stimulate an official investigation; many observers give credit to Daniel Greenberg, editor of *Science & Government Report*, for bringing the local investigation of Stephen Breuning to national political attention.[60] Allan Mazur attributes the same role to the press overall, especially "in precipitating more responsible inquiries by institutions into allegations of misconduct."[61] His study of media reports on eight fraud cases concluded that in five of them the press played a very active role, urging action rather than only reporting on events.

That type of effort was apparent in the Cyril Burt case, where London *Sunday Times* medical correspondent Oliver Gillie apparently "picked up and developed" doubts that had been circulating privately in the scientific community for several years.[62] Gillie's article in the *Times* of 24 October 1976 was reportedly the "first printed accusation" of Burt's fraud and it was Gillie who discovered that Burt's research assistants "probably never existed at all, and certainly had nothing to do with the research he purportedly carried out."[63]

In addition to professional journalists, several other people played out various aspects of the nemesis role in the Burt case. Questions about Burt's work were raised among academics shortly after he had delivered the prestigious Thorndike Award lecture in 1971 to the American Psychological Association. A Princeton student who had heard the lecture brought Burt's work to the attention of psychologist Leon Kamin, who later became "convinced that Burt was a fraud." After some reanalysis of the Burt data, Kamin published a book in 1974 on the science and politics of intelligence testing which included a devastating denunciation of Burt.[64] These criticisms were then picked up by American science writers and by Gillie, who first tried to locate Burt's coauthors, "Miss Margaret Howard" and "Miss J. Conway."[65] Howard and Conway had supposedly helped in the 1950s experiments but Gillie could find no evidence that they ever existed.[66] By 1979, sufficient inconsistencies had been uncovered that a major biography of Burt, originally intended as sympathetic, instead painted a portrait that permanently damaged Burt's reputation.[67]

Constructing the journalists' role as always proactive ignores their other significance as a passive tool. Participants in fraud cases have certainly "used" the press to pressure investigations or simply to publicize their own views. Stewart and Feder mailed copies of their manuscript and their correspondence to journalists as well as scientists; later, they adopted the same techniques to criticize David Baltimore. In the latter case, Baltimore himself turned out to be a skillful manipulator of press coverage; both the biologist and one of his supporters sent open letters to friends throughout the scientific community, encouraging them to write members of Congress and the editors of the major newspapers.[68] Before he testified to Congress in 1989, Baltimore gave private interviews to reporters and accepted an invitation to speak at a dinner meeting of the Washington-area science writers association, held right before the hearing.[69] Theresa Imanishi-Kari held formal press conferences before the 1990 hearings and informal ones in the halls outside the hearing room.[70]

With several notable exceptions, if one looks closely at the role of the press in the most sensational cases, science journalists as a whole appear relatively tame. They frequently jumped on the journalistic bandwagon as it was already rolling down the street, reporting on investigations initiated by NIH or congressional staff, but not breaking new stories on major allegations. This lack of action stands in contrast to the laudable role science journalists have played on other stories, for example, the outstanding reporting on AIDS research.

Unfounded allegations can ruin careers. No one wants journalists to become tabloidistic scandalmongers. Nevertheless, the reporters who know science best do have a special obligation for vigilance. On this issue they must balance their naturally protective attitudes toward science and their empathy for science with the obligation to assist the flow of news in a democratic system. Barbara Yuncker, the journalist who broke the William Summerlin story in 1974, observed later that science reporters routinely must balance the public's right to know against the accused researcher's right to privacy. The Summerlin story first came to her attention as a vague rumor about scandal at Sloan-Kettering; because of her earlier interviews with many of the scientists there, she recognized the research involved and began to investigate. Although the laboratory's administration would have liked to dissuade her from printing the story, they eventually answered all her questions. Failing to inform the press would have been a "serious ethical error" on the part of the laboratory, she later observed, because the public has a right to know how public funds are being used, or misused.[71] The leaks of documents and information associated with more recent cases may present even tougher choices to the journalist who wants to be fair yet share all the news (that's fit to print) with a curious readership.

Future Actors

Focusing on nemesis figures helps to emphasize how essential it is to have impartial procedures at all institutional levels which encourage responsible early reporting of suspicions, enable prompt and thorough investigations, and insure fair treatment of the accused. When systems fail, when investigations drag on for seven or more years, then scientists who are disturbed by what they perceive as unethical practices may grow impatient and attempt to circumvent official rules and take matters into their own hands. Although nemesis figures may eventually provoke appropriate action by officials who should have investigated in the first place, it is their *existence* rather than their particular actions that is the more important issue for science policy. The appearance of more than one nemesis indicates a serious shortcoming in the appropriate channels and indicates probable failure by those managers, administrators, and institutions who should be conducting investigations. Nemesis figures take up a fight when the system seems unwilling or unable to proceed. Moreover, their actions are, by their very nature, amateur efforts, full of passion and outrage but with no authority or official status and no institutional obligation to be fair and just. That characteristic poses risks to the rights of all involved.

I do not wish to imply approval of all actions of nemesis figures or whistleblowers; but likewise, I cannot condemn outright their motives or reject their usefulness. We may simply have to tolerate their presence until universities and research organizations find ways to treat well-intentioned questions about research conduct fairly and seriously. Concerned scientists have stepped into the nemesis role to force ethical responsiveness in seemingly unresponsive institutions. Sometimes, of course, the accuser's motives are not admirable, resting on greed, animosity, or professional jealousy. Any person who constructs a "campaign for justice" behind the scenes by circulating negative letters and repeating gossip deserves tough scrutiny. Nevertheless, the nemesis does the community a service and will continue to do so until the climate for criticism changes. As Walter Gilbert commented:

> There is a canon of the establishment which says that when
> someone objects, that person must be a malcontent and be badly
> motivated and that science is holier than anyone or anything.
> This is the issue: what happens when a scientist is called upon to
> be unsure of his or her work.[72]

One disturbing new trend may force a restructuring of the federal investigating process. Under 1986 amendments to the False Claims Act of 1863, whistleblowers in the United States who allege that a particular institution has allowed fraudulent use of federal grant or contract funds may eventually be awarded up to 30 percent of any money the government recovers. This legislation introduces an unambiguous motive of financial gain. If an individual files a *qui tam* motion in the name of the federal government, the U.S. Department of Justice must then decide whether to pursue a case. In August 1990, the first such charges made under the act, involving accusations against two universities, were made public.[73] Although it is not clear whether the number of allegations will escalate as a result, the increased opportunity for economic reward will put more pressure on authorities to scrutinize carefully the motives of all accusers, even as they develop procedures to ensure due consideration to their claims and appropriate protection of their interests.

Whether we approve of their motives, methods, and missions, whistle-blowers and nemesis figures in science draw attention to deficiencies in the research system and in the policies, public and private, for dealing with problems. It's in the best tradition. American society has long used gadflies and critics as pointers to those problems that need fixing. When editorial or expert review fails to detect plagiarism or error in scientific publications, the nemesis figure can set in motion efforts to ascertain the truth, to locate the wrongdoer, and to rid the literature of the falsity. But a research community that relies primarily on "unofficial spokesmen" to police its members' conformity to ethical norms inexcusably neglects its own responsibility for such monitoring. If nemesis figures are inept or have dishonorable motives, then that neglect may result in terrible injustice to innocent parties. As Daniel S. Greenberg states, when the complaints of whistleblowers are justified, they perform a valuable service to science; but that service must "be looked at against the background of nobody performing it at all when they started out."[74] Far better to create an environment in which "unofficial" or "self-appointed" crimefighters are unneeded.

7 Action

Investigation and Evidence

The steps necessary to complete a successful prosecution for forgery are, first, to prove beyond reasonable doubt that the object is spurious, and then to connect it with the perpetrator.

—George Savage[1]

Given the possession of the primary qualities, namely an open mind and a suspicious nature, there are various ways of dealing with a suspected forgery. One can examine its pedigree and its recent history; one can judge it aesthetically; or one can call in the aid of science.

—Bernard Ashmole[2]

The publishing community's traditional inclination to keep quiet rather than to publicize allegations of plagiarism or fraud attracts criticism from those who regard journals as a Maginot line against unethical behavior. "The timidity of many editors, inclining some . . . to suppress any suggestions of error, minor or major, and even to delete from the indices the entries for major articles to which discreditable suspicions have become attached," one critic writes, compounds the twin problems of investigation and correction.[3]

The quarrel is usually not with the journals' initial choices. Discretion seems prudent during the first moments when a suspicion is voiced, and there is no evidence that more aggressive action then would make much difference in the outcome. Moreover, the risks of hasty action are considerable. Even the briefest investigation can have serious negative consequences for all involved; apart from the waste of time and resources, those who are unfairly or maliciously accused may find their reputations sullied by the resulting publicity, a phenomenon characteristic of all types of sensationalistic charges.[4] The possibility of undesirable consequences, however, should not obscure the positive role that editors can play in initiating or assisting investigations of wrongdoing. Instead, we should take a second look at the journals' internal policies and procedures for handling suspicions. If federal regulations eventually *require* journals to act on all suspicions, then discretion will no longer be an option.

FICTION VERSUS FACT: *THE AFFAIR*

Investigating misconduct that occurs during the publishing process will draw the interests of several different organizations into conflict. Investigations commonly involve a university (sometimes more than one) who employs (or once employed) the accused; if the conduct occurred in the context of a federal grant or contract, then one or more government agencies may be involved. In such instances, quick initial action may be impossible if an agency needs to assemble legal evidence. Diligent investigatory techniques may conflict with the emotionally-charged atmosphere of academic investigations, where personal agendas and intellectual warfare simmer under a thin coating of rhetorical or procedural formality. The rough and tumble rules of the courtroom seem far removed from rarified scholarly discourse, yet the academic discussion may be far less objective under the surface.

Because the records of cases at most American universities are sealed, information of the typical progress of such cases comes primarily from the public statements or testimony of participants. However, the internal conflicts that erupt around allegations of fraud within universities are typical of those that C. P. Snow portrays in his novel of postwar academe in England, *The Affair*, part of his "Strangers and Brothers" series.[5]

In 1953–1954, the central protagonist in the series, barrister Lewis Eliot, works in a high-level government post but is persuaded to assist his Cambridge college as it investigates allegations made about Donald Howard, an untenured research fellow. The episode begins shortly after the death of Professor C. J. B. Palairet, F.R.S., when several American physicists could not replicate results reported in an article that Palairet coauthored with Howard. The Americans called one of the article's diffraction photos "fishy." Closer examination reveals that the illustration has been enlarged and did not come from the experiment described.

At first, Howard denies that the photo is forged; later, he acknowledges forgery but claims that because Palairet supplied the photo, the senior man was therefore the culprit. The college's Court of Seniors rejects this explanation and fires Howard—"Without any noise at all, he had been got rid of." "It was a kind of dismissal" executed only once before in the college's history and so Eliot has assumed that it was done "with . . . obsessive care," and to have been "punctilious and fair."[6]

Howard maintains his innocence. His supporters, convinced that the dismissal represents a case of "sheer blind prejudice" against Howard's leftist sympathies, fight for an appeal and attempt to enlist Eliot's help; but the science faculty vigorously opposes any reopening of the case. They fear that discussion would reflect negatively on science's public image, giving a false impression of scientific dishonesty and damaging the college's reputation. "Considering the chances and the temptation," the number of frauds in science are "astonishingly low," far less than one might expect, one character argues.[7] Many others in the university simply regard Howard as "a twister," "an unmitigated swine" with no redeeming qualities, who is disloyal to the college ("wanting to pull the world down round our ears") and ungrateful to his mentors ("biting the hand that fed him").[8] Palairet, on the other hand, seemingly had little to gain from committing fraud.

The arguments against reopening the case fall apart when a senior faculty member accidentally uncovers new evidence and announces that he now believes Howard's story. In processing the last box of Palairet's papers for the college archives, Julian Skeffington discovers facts that support Howard's statements. As "a brave and honorable man," Skeffington "had not an instant's hesitation, once he believed that Howard was innocent." He ignored personal inconvenience and potential damage to his own career; justice alone seemed important.[9]

The emotions simmering within the college burst to the surface during Christmas night dinner at the Master's lodgings, as the guests (including Eliot and his brother) weigh the rights of the individual (Howard) against the good of the institution. The Master believes that the college—his first concern, his first responsibility—has in fact been "remarkably lucky" in thus far avoiding serious scandal. A "piece of scientific fraud is of course unforgivable," he reflects, but "any unnecessary publicity about it . . . is [also] as near unforgivable as makes no matter."[10] Anyone "who resurrects the trouble is taking a grave responsibility upon himself" because it would achieve nothing except "harm for the college."

As in *Gaudy Night*, the discussants consider the thorny question of whether, once evidence is uncovered, it must be acted upon. One protagonist blurts out to the scientists present: "You'd better face it, this isn't just your own private game." To which one scientist responds: "It might be more convenient if it were."[11] All recognize that a new inquiry will "split the college from top to bottom" and publicity will harm everyone associated with it (one must protect one's career) and "make the place unlivable" (one must preserve the scholarly environment). Secrecy brings other risks, however. If the story should break in the newspapers, then the college might be in "a worse mess than ever" for appearing to have hushed things up.[12] The academics' insecurities (both in personality and employment) are obvious as they flee any hint of discomfort and disruption.

Ultimately, the scientists break the deadlock. Francis Getliffe, an eminent radio astronomer, had initially greeted Skeffington's discovery with hostility. Later, he admits that this hesitation merely reflected his own careerism—his unwillingness to risk making himself "unpleasant to everyone who counts for anything" at the college, "to blot my copy book" or to damage potential support for his candidacy for administrative office. Eventually, Getliffe offers support, although he continues to blame Howard. If Howard is innocent of deliberate deception, Getliffe reasons, then he is guilty of a "greater crime"—stupidity. If Howard "had had the scientific judgment of a newt, he'd never have taken the old man's experiments on trust," Getliffe explains. "It's almost unbelievable that anyone working in that field allowed it [to be published] without having another look."[13]

When asked to be Howard's counsel during the formal reexamination, Eliot hesitates. He is convinced of the scientist's innocence, and his passion for the truth runs deep; but any engagement will disrupt his well-organized, well-protected life. "There is great dignity in being a spectator," he muses.[14] Yet, lack of involvement can also deaden one's conscience: "I had met few people who, made aware beyond all self-deception of an inconvenient fact, were not at its mercy. Hypocrites who saw the naked truth and acted quite contrary—they were a romantic conception. Those whom we call hypocrites simply had a gift for denying to themselves what the truth was."[15]

Eliot becomes an aloof, dignified investigator and advocate, and in this role, he serves as one of several nemesis-like figures in the novel, each with a crucial deficiency. Skeffington recognized the truth in the evidence and blurted out his opinions but lacked the political clout or expertise to force the case. Martin Eliot possessed the political skills, but acted only when it was personally advantageous. Lewis Eliot stands alone in the spotlight, but his status is ambiguous—once a college insider and still loyal to it, he is nevertheless now an outsider to its politics.

Some in the college regard the entire affair as a "squabble among scientists," strongly resenting what they characterize as the scientists' "Pecksniffery."[16] This resentment reflects typical postwar political attitudes to science in Europe and North America, and demonstrates Snow's beliefs about a widening gulf between the "two cultures," separated by morality not methodology. Criticizing the scientific community's Cold

War political maneuvering, one character complains, "You think you can do anything you like with the rest of us, and switch on the moral uplift whenever you feel good."[17]

When the initial bid to reopen the case fails, the college splits further apart. As in other closed communities, the antagonisms fester within the gates. Some of Howard's supporters threaten to "make the whole case public," "to blow the whole thing open" in order to get "simple justice," but even they fear harmful publicity to the college.[18] Finally, the administration agrees to hold another hearing, under the extraordinary stipulation that their ruling will be final and unappealable. Moreover, in a curious twist, Howard must prove his innocence, something he can only do by proving Palairet guilty.

Howard's ultimate vindication represents, therefore, a bitter victory. The reputation of his mentor is damaged. Palairet had been working on an application of his technique, one he expected and wanted to work; the forged photo was intended as confirmation of those expectations. Eliot speculates that Palairet probably acted out of "a curious kind of vanity," as if thinking:

> "I have been right so often. I know I'm right this time. This is the way the world was designed. If the evidence isn't forthcoming, then just for the present I'll produce the evidence. It will show everyone I am right. Then no doubt in the future, others will do experiments and prove how right I was."[19]

In Palairet, modesty mixed with vanity and "perhaps, a spirit of mischief," so that Palairet could believe "this is what I can get away with."[20] As the affair ends, Snow shows how easily an innocent victim—guilty of naiveté, perhaps "stupid" in having accepted evidence without checking it—can himself become the *bête noire*. The college cannot stomach its own failings—although the Fellows restore Howard to his post, they simultaneously deny him renewal and tenure.

In this novel (as in real life), the "deplorable business" of fraud forces the college faculty to question their own assumptions about honesty in their ranks. The novel's outcome, in that respect, is not pleasant. Although investigation seems "intolerable," its resolution also sticks in the craw. To the Master of the college, the "distasteful business of scientific cheating" represents "by its nature, a denial of all that a man of science lives for or ought to live for," but even he realizes that the affair foreshadows events to come. There are "straws in the wind," he observes, "which give reason to think that contemporary standards among a new scientific generation are in a process of decline."[21]

HANDLING SUSPICIONS

The fictional case exposed in *The Affair* may, in fact, be less complicated than many real investigations; many have taken years to complete, involved parallel inquiries by several different research institutions and government organizations, and affected dozens of colleagues and co-workers, journals and publishers. Yet, most cases begin, as did *The Affair*, with a single question raised by a single person, when an editor is contacted by an organization or ethics committee, when staff or editor question a manuscript's authenticity.

Allegations that come to a journal without warning pose the most difficult problems because there is no convenient way to assess quickly an accusation's merit, an accuser's authority or veracity, or even the most appropriate course of action. The timing of a notification, before or after publication, is a crucial factor. In June 1986, allegations were made at the University of Cincinnati about an article that a member of that university's faculty intended to publish in the August 1986 issue of *Pediatrics*, the journal of the American Academy of Pediatrics.[22] The university set up an inquiry in July and, shortly after the article appeared, "wrote the editor to disassociate itself from the article." Commenting later on the subsequent NIH investigation, *Pediatrics* editor Jerold Lucey told a reporter that he was "annoyed" at NIH and the university for not having notified him earlier that the paper was being questioned; he pointed out that the "flaws were not of the type that could have been detected in the peer review process" and he might have chosen to delay publication until the matter was resolved.[23]

In another case, a university history department concluded that a junior member of their faculty had plagiarized much of a book manuscript then currently in press; but the department decided not to notify the publisher.[24] The historians later explained that they had feared the potential legal repercussions of such notification. No vote had actually taken place on the accused scholar's tenure case; the university had issued no public statement and mounted no official investigation; instead, he had been allowed to resign quietly. Even after the book was published, neither the author nor the publisher of the plagiarized material sued for copyright violation, although the offended author did request a formal inquiry by the American Association of University Professors.[25] Thomas Mallon concludes about the case that many of the people involved, perhaps out of timidity or cupidity, failed to look beyond their own self-interest or beyond what they perceived to be the best interests of their institution. They appeared to disavow accountability to any other person or institution or even to the discipline of history or the integrity of its scholarly literature.

Hopeful signs have appeared in more recent investigations. In one case, journals that had published questioned work were notified about allegations at the same time as the government agency that had funded the research, and the journals were kept informed of progress throughout the formal investigation.[26]

Because of the chance that accusations may be false or spurious, editors have tended to proceed with caution. Sometimes there is a reasonable explanation for what is initially perceived as misconduct; sometimes, "plagiarism" may be inadvertent or unintentional. On occasion, allegations have involved only relatively minor aspects of manuscripts albeit ones that seem crucial to the accusers. Territorial disputes between collaborators (or former collaborators) or rivals, and scholarly arguments over facts or interpretations, can be disguised as accusations of misconduct. A few "ethics" disputes have actually involved legitimate disagreements in scholarly interpretation, of the type formerly played out in letters columns or in seminars.

Editors also attract gossip (and some encourage it), so the original source of a secondhand allegation may be in doubt. Benjamin Lewin, editor of *Cell*, has described one fraud case that was uncovered only after gossip in the field grew strong—in the form of "doubts expressed after presentation of work at a meeting."[27] Those who disregard rumor and then find themselves victims of deception may wish they had acted otherwise, but triggering an ethics investigation without corroborating evidence could inadvertently damage an innocent person's career. The environment surrounding allegations can be so full of emotion that those who feel responsible in some way (perhaps because they failed to see what was happening), even if they are not accused of wrongdoing themselves, may suffer from guilt and remorse; in one case, suicide was the terrible result.[28]

When allegations are made, it is best to have some consistent policy for handling them. According to Lewin, *Cell* has tried various approaches— some not so successful—to handling suspicions.[29] In one case, allegations circulated as gossip in the field, but the journal had no proof; *Cell* published the article and it was later proved to be a fraud. In another case, a referee pointed out a problem and the authors were duly notified; the senior author then rechecked the work and withdrew the article. A third article was published, despite the journal's suspicions at the time of submission; the author later admitted forgery of other papers but not of this particular article and it remains in the literature. Lewin believes that, in each case, the journal made an appropriate "presumption of innocence" and that such a presumption is necessary for journals to carry out their work.

Motivating an Investigation

Once a journal decides to act on a suspicion beyond a quick check of its own files, then a number of important decisions must be made. Should it proceed with a quiet inquiry on its own, or notify the accused's employer and any government agency that funded the research and let them conduct an investigation? Some editors advocate a "dynamic" two-step approach, suggesting that an author (or referee) first be notified of the questions; if the author's explanation does not seem satisfactory, then the editor relays the suspicions to the author's employer.[30] Today, a heightened fear of litigation discourages some from taking any action beyond simply notifying the authors that questions have been raised; but, whenever feasible and appropriate, most commentators on this issue advise that the journal should first try to persuade an accuser to raise the allegations directly to the employer or sponsoring institution of the accused.

The pressure for journals to initiate their own investigations stems in part from disappointment in how research institutions, notably universities, have handled cases. Universities at first attempted to keep incidents quiet and to avoid calling in government investigators; they also often failed to notify journals, either when allegations were being investigated or after decisions were reached. A university issued a discreet announcement of the results of its inquiry (usually after the person was dismissed), but, in the meantime, questionable articles had moved through the publication process. In the John Darsee case, for example, Harvard University did not rush to notify the federal sponsors of Darsee's research and, despite an increasingly grave situation, did "nothing to make it public."[31] The assertion of institutional foot-dragging by the university was later used effectively in congressional hearings to justify the need for federal regulations.

Understandably, editors may feel reluctant to notify an employer because charges may later prove to be false, yet no one wants to see genuine malfeasance or unethical conduct continue unrestrained. A partial solution to this dilemma is establishment of an ombudsman or ombudswoman, an administrative officer to whom students, faculty, and employees can confidentially report instances of racial and sexual harassment, biased employment practices, and complaints about fraud, waste, or abuse. These intermediaries can also assist in preliminary investigations of scientific misconduct and could be asked to intervene in a dispute between a university employee and a journal. Other alternative solutions are the "fraud hotlines" set up at government agencies to facilitate reporting of all types of fraud, waste, and abuse and the ethics officers or inspectors general appointed at universities and federal agencies. Finally, universities receiving federal funding now are appointing institutional ethics committees to investigate research misconduct allegations and such committees would often be the best place for an editor or publisher to direct questions about these matters.

Private Inquiries

Not everyone is content to allow problems to be solved "elsewhere." Journals have in fact launched their own independent investigations, perhaps the most controversial of which took place in 1988. That July, the journal *Nature* announced that it was publishing a paper that both editor John Maddox and the manuscript's referees had found "unbelievable." The journal also announced that it had required, as a condition of publication, that the author repeat his experiments while being observed by an "investigative team" appointed and funded by the journal.[32] Four weeks later, the *Nature* team reported that the research results were "a delusion" and "not reproducible."[33]

The paper in question had first been submitted to *Nature* in 1986 by Jacques Benveniste, a professor at the University of Paris (Sud) and research director at the National Institute of Health and Medical Research (INSERM) in France. His findings suggested that "an antibody returns to its biological effect even when aqueous solutions of it are diluted to the point where no molecules of the antibody can reasonably be expected to remain."[34] Because this conclusion was widely regarded as offering support for the practices of homeopathic medicine, Benveniste's work had drawn considerable criticism from more conservative biologists. Nevertheless, after extensive review, referees had found no "obvious flaws" in the manuscript and the experiment had been repeated in five laboratories around the world.

John Maddox remained skeptical, so he convinced Benveniste to accept *Nature's* conditions for publication. Maddox also added an "editorial note" to the article, published a separate editorial expressing his skepticism, and announced the existence of the investigating team, which consisted of Maddox, NIH biologist Walter Stewart, and magician James ("The Amazing") Randi. Their report, which appeared in the 28 July 1988 *Nature*, alleged errors in Benveniste's procedures and alleged that he had failed to review contradictory evidence, but it did not charge intentional fraud. Benveniste angrily rebutted their criticism in the same issue and in letters to other journals, charging that the *Nature* investigation had been "a mockery of scientific inquiry" and that the report was full of inaccuracies. He later commented that, although he had welcomed the opportunity to explain his results, he felt "betrayed" by Maddox's decision to publish the article apparently just to attack it.[35]

By September 1988, the episode had become, as one news magazine tagged it, *l'affaire Benveniste*.[36] There was considerable criticism of *Nature* for publishing the article even though questions about it remained unresolved. Maddox pleaded that he had succumbed to "a mixture of exasperation and curiosity"; the author had badgered him constantly and the French press had begun to "herald the findings" as if they were already verified.[37] Maddox's critics argued that he might have waited another four weeks while the investigation was completed. Maddox countered that he had been afraid that Benveniste would withdraw the paper upon seeing the team's report.[38]

Nature clearly departed from the high road that scientific journals traditionally claim as their exclusive territory, but both supporters and detractors of Benveniste found fault with the process, and no clear agreement emerged as to what *should* have been the proper course of action. Geneticist Norton Zinder wrote that "*Nature* no longer behaves like a serious scientific journal," and he complained that the journal's actions gave Benveniste "an out" to criticism of his work.[39] Sociologist Harry Collins called the episode a "circus," one that would eventually be "bad" for science's public image.[40] Bernard Dixon, a journalist who has covered the issue extensively, was more supportive of Maddox's strategy: "Journals should actively investigate bizarre claims before deciding to publish," lest the general public view come to regard publication as an "imprimatur" of legitimacy.[41] Publisher Eugene Garfield, too, appeared to support the idea of some type of investigation before publication as a form of extended, "intensive" peer review, but he nevertheless faulted *Nature* for "poor judgment" in sending an unqualified team to investigate *after* publication, a process he regarded as sensationalistic.[42]

The harshest criticism centered on whether Maddox had exceeded the limits of an editor's responsibility. Arnold Relman of *NEJM* wondered why Maddox had not insisted on independent verification before publishing. "A journal should not be an investigative body," Relman remarked; an editor's role is to ensure fair and rigorous review. When "the editor becomes the judge, the jury, the plaintiff and—in some sense—the accused," then there is a serious conflict of interest.[43] Maddox, he said, was "seriously mistaken" in suggesting that editors investigate the authenticity of manuscripts; neither editors nor their reviewers have the resources or the authority "to act as laboratory cops."[44] "On-site verification of submitted work," even conducted sporadically, could destroy the "presumption of trust in the honesty (but not the infallibility) of one's colleagues" that is so vital to science as well as to the editor-author relationship. Said Relman: "If an editor has good reason to suspect fraud, and not simply error, he has an obligation to see that an investigation is launched through appropriate channels; even in such cases he should not, and cannot, be the investigator himself." That responsibility, Relman concluded, "belongs primarily with the institution sponsoring the investigator's work."[45]

In the debate over *l'affaire Benveniste*, many commentators focused on the *pace* of the editorial process as among the most problematical aspects. Confirmation or refutation of questionable text or data, they argued, should not be hurried because speed can confound good judgment. Many months later, Benveniste's own institution, INSERM, decided to allow him to continue the research in order to confirm his earlier results. The institution's official statement cited the *Nature* investigation as one of the reasons for this decision—stating that the controversy had been fueled by what INSERM called "questionable ulterior justifications."[46]

This episode, it turns out, was not the first in which *Nature* had arranged an independent investigation or had run extensive criticism alongside an article. In one of the three previous instances, however, the process had resulted in *Nature*'s decision *not* to publish an article.

In a quite different situation, private inquiries will be necessary: when allegations involve an editor, consulting editor, staff member, or referee.[47] If the journal is sponsored or owned by a professional association, allegations may be referred to its standing committees on ethics or professional standards. If a journal is independent, or owned by a university or commercial press, then investigations will most likely be affected by the publisher's policy, editor's contract, or the rules of the accused's employer.

HANDLING EVIDENCE

Once any organization—small, independent journal or major university—decides to investigate, it immediately confronts questions of *what* type of evidence will be required to prove or disprove the allegation, *whether* such evidence exists and can be obtained, and *who* should be required to supply or obtain it.

Plagiarism accusations normally present fewer problems than accusations of fabricated or forged data. Many plagiarisms are detectable by sharp eyesight, good memory, and intellectual alertness. If side-by-side comparisons of two texts show that half of the paragraphs of one are identical or very similar to paragraphs in the other *and* if the "original" text is not cited, then plagiarism has probably occurred. The use of new computer "tests" for plagiarism, which match combinations of words and phrases, are of questionable usefulness at this point because their track record in demonstrating deliberate plagiarism in technical, as opposed to literary, text is limited. Whatever the test used, however, common sense dictates that corroborating evidence be obtained—for example, that an alleged plagiarist had had access to the original text, especially if the original text is a proposal, student paper, or similar unpublished document.

One of the most crucial questions is who bears the burden of proof for innocence or guilt, because that decision will determine the type of evidence required and the level of involvement expected of a journal. There is little agreement on this point, probably because it requires adoption of the rules of law rather than of science. In a statement typical of those attitudes, one critic has argued strongly that the burden of proof should be on the *scientist* to prove he is correct, not on others to prove he is wrong.[48] A scientist who has been accused of misconduct has responded to that argument by noting that in other "public accusation[s] of fraud, the burden of proof lies on the accuser"; moreover, "should the accusation prove false . . . the individual responsible for the inflammatory statements should be held accountable."[49]

Investigations of research misconduct conducted by some universities and government agencies unfortunately still appear to be insensitive to the rights of the accused and still seek to force the accused, or the accused's supporters, to prove innocence. That approach mimics the assumptions of science itself, where a researcher provides evidence of having reached the goal described, or of having proven the hypothesis.[50] Journalist Roger Lewin has proposed that in misconduct inquiries "the burden of proof should be on the laboratory: if any work in which a forger has had a hand cannot be judged authentic on the basis of detailed examination of the original data, it should be retracted."[51] Certainly, science requires that proponents of radical new theories or interpretations "prove" why conventional wisdom should be overruled, but adopting the same requirement for misconduct allegations could unfairly burden an accused person and multiply the negative effects of false accusations.

The potential array and type of evidence which can be assembled by investigators will ultimately shape the type of investigation. To prove that data or artifacts have been fabricated, various types of evidence are being used, some borrowing techniques from those used in detecting forgery in other scholarly and creative fields, such as combining subjective judgment with physical evidence. In detecting art forgery, for example, it is said to be crucial to coordinate connoisseurship with technical tests, to supplement "the eye with careful research and scientific examination."[52] To assess originality of literary authorship, scholars of literature combine internal and external evidence. Internal evidence consists of data from such techniques as textual analysis, which compares parallel passages, style, or the order and arrangement of ideas. It might also include the use of computer-matching programs for matching texts. External evidence includes the history of a questioned data book (e.g., who created it, or who had possession of it and when), or handwriting or paper analysis. When a possible forged object is at issue, examination may focus on material or workmanship for clues about its origin or manufacture.[53]

A fascinating example of the use of such combined evidence, and one that influenced its wider development and use in determining authorship, took place in the late 1940s when historians began to question authenticity of a famous three-volume work, *The Horn Papers: Early Westward Movement on the Monongahela and Upper Ohio, 1765–1795*, published by an Ohio historical society.[54] The volumes had appeared to be an extraordinary find, because they included diaries, court dockets, maps, and other papers of early pioneers, one of whose descendants, Joseph Horn, had edited and published the papers. When the historian Julian P. Boyd first charged in 1946 that *The Horn Papers* were forged, many of his colleagues vehemently disagreed. Boyd's suspicions had first been raised by internal evidence (in this instance, certain inconsistencies in style and historical inaccuracies), but the rather substantial amount of external evidence—that is, "the mere bulk and complexity of the collection of papers, the impressive numbers of collateral artifacts, the absence of any pecuniary or other compelling motive for forgery, the unquestioned sincerity of the sponsoring society, the many statements in the manuscripts that agree with generally accepted facts"—favored authenticity.[55] Eventually, technically based external evidence confirmed forgery. Documents supporting *The Horn Papers* were sent to the National Archives in Washington, and chemical and other tests showed them to be a clever pastiche. In retrospect, the signs of fakery had been abundant early on—the supporting artifacts were not authentic (e.g., the ink and paper were artificially aged) and there were crucial anachronisms in the internal and external evidence; the substance of *The Horn Papers*, however, had seemed convincing evidence of their genuineness.[56] Only close scrutiny of individual documents apart from that context allowed sufficient objectivity to spot the deliberate fraud.

Those who suspect fraud in the context of research communication may confront similar contradictory evidence, especially in the description of skillfully forged or fabricated data for several research projects. Apparent inconsistencies within articles may be initially outweighed by: (1) the extent of the research base; (2) the absence of any pecuniary motive; (3) the authority and reputation of the accused's employer or supervisor; (4) the appearance of the fraud in the context of well-established theory and practice; or (5) the reputation of the accused.

Inconsistencies in internal evidence stimulated the initial questioning of the work of Cyril Burt; but here, too, Burt's contributions to science had shielded him from criticism and made deception seem improbable. Eventually, some skeptics questioned why the statistical correlations used to support conclusions in papers he published from 1913 to 1966 remained not just consistent but identical, even though his project supposedly included an increasing number of experimental subjects. This fact, as journalist Nigel Hawkes observes, was "a mathematical impossibility which seems to have gone unnoticed at the time [of initial publication]" but was noticeable when scientists examined the work as a whole.[57] After analyzing Burt's personal diaries, one of his biographers concludes that, because the diaries were exceptionally detailed but omitted discussion of certain key research projects, Burt may not have conducted those projects at all.[58]

Historians William M. Calder and David A. Traill have used Heinrich Schliemann's letters and diaries to demonstrate a similar deception—that he probably did not find the famous "Treasure of Priam," one of the most famous archaeological discoveries of the nineteenth century, as he had claimed.[59] Calder notes that Schliemann's deception was made easier by the looseness of archaeological techniques and practices at the time:

> Archeology is the only discipline in ancient studies which to attain its ends is required to destroy its evidence. Once excavated, a site must be reconstructed from literary evidence, letters, notebooks, diaries, and published reports, aided, if available, by drawings and photographs.[60]

Using just such evidence, the two historians have shown that Schliemann could not have been present at the relevant dig during critical periods, that he had had exceptionally cozy relationships with dealers and others who could have helped to fabricate (or supply) artifacts for a "Trojan treasure," and that there was strong evidence of a pattern of prevarication throughout his life (including a famous but indisputably spurious "eyewitness account" of the San Francisco Fire of 1851).

Each of these cases represents the application of investigative techniques to the work of people long dead, unable to defend themselves or sue for libel if allegations prove false. Applying technical tests in contemporary cases will necessarily be more problematic because it will require the cooperation of all scientists involved, including the accused; moreover, access to and preservation of evidence is crucial for legal prosecution of a case.

Without a confession, it may be exceedingly difficult to prove intentional deception or biased reporting of data.[61] Suspicions may be voiced but proof will be elusive, especially in the social sciences, where "facts" supporting conclusions "can bear more than a single interpretation" or in fields where paradigms or general theories are undergoing shift or reversal.[62] As historian Bernard Ashmole observes, aesthetic judgment about the truth or untruth of a work must be part of any initial assessment. Aesthetic judgment alone cannot provide sufficient evidence to prove forgery or deception because such judgment can be fallible and is "bound to be subjective"; yet it must not be ignored altogether because it exploits the wisdom and experience of the expert investigator.[63] In the context of deception by scientists, what we might call "scientific judgment" or intuition plays a similar role.

The provenance of a document or object provides additional clues to its legitimacy. In detecting forgery of art or artifacts, experts ask whether an object's origins seem suspicious; has it been discovered "in doubtful company" or can every step of its existence be traced?[64] In science, attempts to fake or alter a provenance might result in laboratory data books (what one might call the "history" of the experiment) falsified after the fact. Such evidence is attracting new attention. During the investigation of David Bridges (see chapter 5), the NIH investigating committee could not obtain the original lab notebooks they needed to trace the project's history.[65] To provide independent corroboration of whether the research in question had taken place at a certain time and in the manner described in the published report, they looked at "discrepancies" between available lab records and the descriptions published in *Science*; they examined Bridges's earlier work for indications of ideas that could have led to the procedures described in the article; and they scrutinized the project's chronology, asking Bridges to certify when specific experiments were performed. His coauthor, for example, told the committee that the experiment had begun in August, rather than in May as Bridges had claimed; this discrepancy placed the probable date of the first experiment at a time after Bridges had reviewed the other group's manuscript. The committee then obtained records of the chemicals purchased for the laboratory, finding that "ordering records and technical information from the supplier" showed that the laboratory did not receive sufficient quantities of a chemical crucial to the experiment until much later than Bridges claimed to have begun the experiment.[66]

For its investigation of immunologist Thereza Imanishi-Kari, the House Committee on Energy and Commerce's Oversight and Investigations Subcommittee secured technical analysis of her notebooks, asking the U.S. Secret Service to analyze the ink, paper, and age of key documents and of the paper tapes from gamma radiation counters used in her experiments.[67] Forensic analysis techniques similar to those used to detect counterfeit currency were applied to notebook pages and printout tapes. Imanishi-Kari had previously testified that she had generated one set of data (which was unpublished but crucial to the authenticity and accuracy of a published paper then in question) in June 1985, but the Secret Service found what they testified were crucial differences, in color, font, and ink, between the printouts they examined and authentic tapes from that time.[68] This type of examination of evidence stands out because the Secret Service conducted its inquiries independent from study of the scientific content of the pages in question; the agents are skilled in document analysis, the scientific analysis of paper and ink, not in immunology or the specific topic of Imanishi-Kari's research.

PROTECTING INDIVIDUAL RIGHTS

Max Friedländer once observed that "it may be an error to buy a work of art and discover that it is a fake, but it is a sin to call a fake something that is genuine"—a sentiment that, when applied to a scientific article, most scientists and editors would probably endorse as well.[69] Of growing importance in the discussion of research fraud is how to improve detection, insure thorough investigation, yet provide due process and protections to individuals accused of wrongdoing. These goals are especially difficult to achieve when investigations may be conducted simultaneously by journals, research institutions, and government agencies and when not only employers and former employers may be involved but also Congress and the courts.

In any investigation of wrongdoing, the rights of several individuals and groups are likely to conflict, but regulations have often been designed to protect institutions not the interests of individuals. Especially when the whistleblower or witness is a student or employee without tenure, prestige, or power, the act of accusation can carry potentially harsh repercussions, both personal and professional, yet the arguments against allowing anonymous whistleblowing are strong, usually revolving around concern that anonymity would encourage "Star Chamber" proceedings.[70] Bridges's attorney raised just such objections during the NIH investigation of his work, asserting that Bridges was deprived of due process because "testimony was not sworn, no transcript was made of the proceedings, and [he] was not allowed to confront his accusers."[71] Although the committee and the NIH denied the charges, the compliant raised important issues about the fairness of investigations, both public and private.

Academics have been notoriously reluctant to apply due process procedures in their investigations—for several reasons. Sometimes the worst side of human nature surfaces; sometimes committees just want to get back to their laboratories or classrooms, where they are not bothered by bureaucratic red tape. One recent controversy involving American historians provides sobering lessons about just this point. In 1981, a rising young historian, David Abraham, published his first book to generally favorable reviews. A few years later, a group of senior historians who disagreed with Abraham's interpretations, began to circulate "Dear Colleague" letters, accusing Abraham of "*mis*interpretation"; they also published letters containing similar allegations in some of the major history journals.[72] The historians allied against Abraham then attempted to persuade Princeton University Press to withdraw Abraham's book and the University of Chicago to rescind his doctoral degree. Although unsuccessful, the efforts coincided with Abraham's attempt to secure his next academic job, and that search was seriously hobbled. Although his defenders tried to persuade the American Historical Association to initiate a formal investigation of either Abraham or his accusers, the association refused. Eventually, as Peter Novick concludes in his account of the case, the accusers "succeeded in their avowed aim of driving Abraham out of the historical profession."[73]

Even to the reader unfamiliar with the specific allegations made about Abraham's work, the case should demonstrate what can happen when a field lacks procedures for initiating objective investigation by disinterested parties and when, instead, allegations are never formally placed on the record, where the accused could respond or explain. Throughout the five-year struggle and despite his repeated attempts to do so, Abraham was never able to confront his accusers directly in any common forum. Although the initial charges centered on disagreement over historical interpretation, the dispute devolved into vague accusations of deliberate distortion, which Abraham could only attempt to refute after the fact. Abraham's critics initially characterized the book as "unworthy of serious consideration," but these assessments, too, shifted over time. No formal complaint was ever made in a forum (such as a university-based ethics committee) in which both sides could present a case, where evidence could be examined by independent investigators, and where Abraham could accept or rebut the accusations. He could only write his own letters for publication in subsequent issues of the same journals his opponents had already exploited. By the time Abraham replied in a later issue to one accusing letter, another letter with new allegations was published elsewhere. Both accusers and defenders tended to react with "outrage and incredulity," Novick concludes, and this emotionally charged climate further exacerbated the dispute.

For those who regard this particular controversy as a test of the history profession's reputation for objectivity—regardless of their convictions about Abraham's guilt or innocence—the outcome was unsatisfying. All participants, and the profession as a whole, suffered because of the ambiguity and nebulous nature of the charges and the absence of procedures allowing resolution. Early investigation might have resolved the dispute, aired the central issues of concern to the profession, and provided due process to the accused. Because no evidence of deliberate unethical conduct was ever presented on the record, the outcome seems especially tragic and distasteful. Gossip and innuendo ruled the process, not fairness or objectivity.

The dilemma of how to ensure fair, equitable administrative or institutional hearings will be with us for some time.[74] At least now when an organization pleads discretion, confidentiality, or conflict of interest as the excuse for suspending due process, that excuse is scrutinized carefully. Journals and publishers, however, have yet to consider how they can promise due process for the accused yet maintain the confidentiality of the review system. What obligations do journal editors have to cooperate with other organizations in providing access to publication records or in testifying about the review process? What role should journals play in the setting and dissemination of mandatory standards?

Although *uniform* standards, procedures, and policies are impractical, some type of guidelines that discourage journals from participating in or assisting "prosecution by innuendo" seems a reasonable first step. Every journal should consider now who it will notify, and when, if allegations are made about its content or review process. Every journal should consider whether it will attempt preliminary investigation, or will automatically transfer responsibility to another official or institution—and, if the latter, which one. Every journal should consider the limits of the responsibility it will accept for prevention, investigation, and (as the next chapter discusses) resolution. It might also anticipate how economics or time will influence its ability to carry out that responsibility. And every journal should consider to what extent it will cooperate with attempts by other institutions or organizations to resolve disputes over research communication.

8 Resolution
Correction, Retraction, Punishment

To put one's theft into print is to have it forever on the library shelves, guiltily stacked just an aisle away from the volume it victimized, a stain that doesn't wash but forever circulates.

—Thomas Mallon[1]

In R. T. Campbell's *Unholy Dying: A Detective Story* (1945), the murder victim, geneticist Ian Porter, is clearly a rotter. When students and postdoctoral assistants work in this fictional scientist's laboratory, they are expected to allow him to "use" their ideas. "The man lives by what he steals," one of Porter's co-workers observes; "he's never stolen anything from me, except my best remarks, and I can't really object to that, though I do get annoyed when he spoils them. But he steals from his assistants and . . . his assistants are under him and they will remain anonymous just as long as he can keep them anonymous."[2] One of Porter's research assistants just as strongly defends the scientist:

> They accused him of stealing the ideas of better men when he suggested some line of work to someone and then carried it to a successful conclusion after the other had failed. Was that stealing? No, I tell you. He was taking what was rightfully his, and no one could disprove his claims to his own property.[3]

Another assistant, however, was not so understanding or forgiving, and arranged Porter's murder.

That so many novelists through the years have chosen forgery and plagiarism as motives for murder reflects the depth of emotions revealed during investigations of real cases. That same intense emotional involvement can also prolong a controversy if one party or another remains unsatisfied with the outcome. Harold Edgar, professor of law at Columbia University, has pointed out that achieving resolution or closure of a case poses one of the greatest problems in handling misconduct allegations today.[4] It appears to make little difference whether a decision is definitive or ambiguous. Even after an institutional ethics committee or administrative officer issues a clear finding of guilt or innocence, supporters on one side or another may still be outraged at what they perceive as injustice and may be unwilling to accept the finding. Sometimes, they begin to press their views through the mass media. An ambiguous decision creates still other problems, for it leaves aggrieved parties on both sides and fails to dispel the cloud of doubt over the accused. If colleagues and co-workers find interactions with the accused to be strained and awkward, then it may be impossible to revive a stalled career. Ambiguity may also reactivate the gossip network as vengeful opponents choose rumor as an alternative to the punishment they believe is now unobtainable through regular channels.

Journals, willingly or not, often become the arenas in which this next act is played out, as committees and institutions attempt to publicize verdicts, correct wrongdoing, and punish the guilty, or as vengeful opponents continue their attacks. Because they are perceived as the "bulletin boards" for news in their fields, journals may be drawn into disputes in which they played no prior role (e.g., as the place where the deception was published or as the original source of plagiarized material).

The most common notification activities include the publication of apologies, corrections, or retractions and the discussion or news analysis that gives publicity to an investigation's outcome. One of these means of resolution—to request or require published correction (either through a retraction letter, notice of error, retraction notice, or withdrawal of a book or journal issue)—has stimulated an array of new legal and ethical dilemmas for publishing organizations. As a result, many journals and their publishers delay action until an accused author confesses or until some type of legal judgment is rendered; some are even refusing to act after official investigations are complete, for fear of being sued for libel by parties who continue to maintain their innocence.

SAYING YOU'RE SORRY: PUBLIC APOLOGIES

The mildest form of resolution to any awkward social situation is usually a simple apology, publicly acknowledging one's error. As sociologists point out, sometimes "it suffices for the offender merely to acknowledge some minor transgression. . . . [A]cknowledgment is a ritual that repairs a disturbed situation."[5] When we yawn, we say, "excuse me." Usually, the people around us accept our apology and do not regard the yawn as deliberate insult. Apologies serve a similar function in science, affording a traditional resolution for embarrassment, as when an author inadvertently makes statistical errors, innocently forgets a citation, or inadvertently omits a coauthor's name. Ritual apologies have also been used to resolve more serious offenses and are all the more effective, John Maddox asserts, when they express well-mannered humor and graciousness.[6]

When a violation is deliberate or extensive, when it falls among the acts considered to be egregious violations of group norms, or when it violates specific legal or administrative regulations, then an apology may seem insufficient restitution, even if cheaper, easier, and more civilized than most alternatives. Plagiarism provides the most problematic example. As Thomas Mallon astutely observes, the plagiarized text and the original text continue to coexist in the literature of a field; they may sit side by side on a library shelf. In that case, apology seems a wan substitute for lost credit, especially if a plagiarist exploited another's words or ideas to achieve a significant publication or financial reward.[7]

Apologies do offer attractive solutions. In what is probably the most celebrated scientific apology to date, David Baltimore in 1991 reversed his previous defense of Thereza Imanishi-Kari and his criticisms of Rep. John Dingell and conceded that Congress did have a responsibility to oversee and investigate federal R&D and that he had failed to investigate adequately the allegations of Margot O'Toole.[8] This particular case is unusual in that Baltimore had served as a lightning rod for the publicity surrounding the case such that it became quickly known in the press as "the Baltimore case" even though the specific allegation of impropriety involved Imanishi-Kari. In other cases involving similar allegations, the name of the accused has usually been the news tag (e.g., "the Darsee case"), but in this instance perhaps the phrase was not a misnomer. The linkage expressed private criticisms and served as a public reprimand to Baltimore for how he had acted toward O'Toole. In his formal letter to the NIH, Baltimore seemed to acknowledge that he had been faulted for these attitudes ("I commend Dr. O'Toole for her courage . . ." and "I regret and apologize to her . . .") and, in the final sentence, reveals that he understands the criticism ("this entire episode has reminded me of the importance of humility in the face of scientific data").[9]

When more than one aggrieved party is involved, resolution may not be so swift—especially if participants cannot agree on an appropriate mode of resolution. Attention begins to focus not on procedure but on damaged honor, not on forgiveness but on revenge. If the complainant perceives the miscreant as unrepentantly arrogant, or if the accused has greater stature in the field, then the injured party's hurt may be especially deep. In such circumstances, resolution will probably require more than saying "I'm sorry," but that should not deter an editor from suggesting (or urging) an offender from offering it.

If the offense involves fabricated data or forged objects, then to whom should the accused apologize—to the readers, the editor, or the intellectual community in general? Despite thorny issues like these, there may still be times when old-fashioned manners can construct a face-saving (and time-saving) solution, if no one has been harmed, no state or federal funds are involved, no laws have been broken, and the accused is genuinely repentant.

RETRACTION

We have all experienced the desire to retrieve a statement or action and erase it from existence, but only cartoon characters can truly savor that opportunity. Usually we correct our mistakes through subsequent statements or actions. We write a second letter to express our regret for one written in haste, for example; we do not invade the addressee's office and steal back the offending missive. A museum that discovers it has purchased a fake does not necessarily throw it away; instead, the object is "de-accessioned," relabeled, or perhaps removed from display. Its status is altered even though it may remain in the museum's storehouse.

In publishing, similar combinations of physical removal and alteration of status may occur. When Johann Bartholomew Adam Beringer discovered that he had based his *Lithographiae Wirceburgensis* (1726) on false data, he attempted to buy back all the copies of his book, an effort that led to his financial ruin.[10] In modern times, publishers formally withdraw books that appear to be libelous or based on falsified or plagiarized material, but cannot usually buy back all existing copies. The responses of journal publishers are different because the physical problem is more difficult. Whether a questionable research article is physically removed from an issue in process but not yet officially published, depends on when questions are raised, on the stage of publication and the costs of withdrawal. Deciding not to publish an article at any time before makeup and typesetting wastes the editorial resources and the time of those involved in review. Once an issue is typeset, printed, or bound, however, the costs become more substantial. To reprint an entire issue without the offending material would represent a prohibitive expense for most journals; an article represents only *part* of the contents of an issue yet all would have to be reprinted.

Most discoveries of forgery or plagiarism occur months or years after publication. In these circumstances, the usual reaction is to publish what is called a "retraction." The term is actually a misnomer because the article is only "retracted" in spirit, not destroyed or physically removed from existing copies of the journal. Instead, a formal notice readjusts the article's status.

Journals willingly publish retractions primarily to maintain their own reputations for accuracy, but retractions also reflect a philosophical belief in the importance of the field's intellectual integrity. Scholars seek to "get rid of forgeries," for example, because existence of spurious artifacts or specimens "corrupts" the body of knowledge that defines a field's substance. In art, forgeries create misunderstandings of an artist's or sculptor's genuine oeuvre; in science, they may lend support to false theories or route causative explanations down the wrong path.[11] Spurious art objects can be removed from galleries, but physical retraction of the scientific knowledge represented in a published journal article is impossible. When copyright violation, plagiarism, or libel occurs, publishers do recall books, even to the extent of repurchasing them from buyers. In 1972, for example, a large U.S. publisher withdrew and destroyed three thousand copies of a novel that apparently contained two pages of verbatim plagiarism.[12] Even if every retail store could trace every purchaser of a particular edition, customers cannot be forced to return books; cooperation in any "buy-back" program is voluntary. As a result, many "withdrawn" books are still widely available. For example, I have located copies of one famous "withdrawn" book (Clifford Irving's *Fake!*[13]) in used bookstores and charity book sales from Seattle to Washington, D.C., and the famous Shakespeare & Co. bookstore on the Left Bank of Paris.

Retraction offers a practical solution to the dilemma. A notice of retraction, published by the original author(s) in the same journal that published his(their) article, remains the most straightforward way to deal with correcting innocent mistakes, technical errors, and deliberate deception.[14] A notice states that the article, its conclusions and/or data, should no longer be regarded as part of the legitimate body of knowledge in the field. Sometimes the notice explains the origin of the error or problem in detail; sometimes it only informs readers that an experiment is being redone or that an article's conclusions are in doubt.

Just a few decades ago, journals tended to use ambiguous phrases to signal problems in which malfeasance was suspected. The statement came from the editor or publications board, not the author, and the language was measured and restrained compared to notices today. In 1961, for example, *Science*'s editorial board chose to describe a questioned article as "an unfortunate event" and to apologize to readers for erroneous illustrations in the article.[15] The editorial board indicated that it was following the lead of another journal in which the same coauthor had published; in the other case, "at least five of the six figures used to document an article . . . came from 'the previously published work of other authors'."[16] In effect, the *Science* board used another journal's actions to validate their own. The tone of the *Science* notice was remote but polite. The list of "discrepancies" (which apparently involved plagiarized illustrations) was straightforward. "We have had our attention drawn to several discrepancies in a paper," the note stated; but no information was given about the specific charges, who had brought them, and when.[17]

The tone of retractions has changed through the years, generally toward more blunt language and toward the direct involvement of authors and coauthors. For example, Efraim Racker, writing to *Science* in 1983, adopted a forthright approach as he reported on which portions of his research with Mark Spector (retracted in 1981) had been confirmed as accurate and which had been found to be fabricated. He never used the words "fraud" or "misconduct" but did recount precisely what he had originally believed Spector to have done, and what subsequent investigation showed Spector had (and had not) done, adding his own speculations about why Spector fabricated the research.[18]

Insufficient data exist to say whether the relative number of retractions is increasing. We also do not know what proportion of the articles being retracted have been questioned as part of some official investigation by university or government offices, or what proportion represent questions raised by coauthors, readers, or editors. The absolute number does not appear to be large, however. In checking its records, the *Annals of Internal Medicine* found that between 1980 and 1985, it published 144 corrections of scientific information in previously published articles, out of more than 6,000 pages published during that period.[19] One-third of these notices dealt with factual errors; another 28 percent described information that had been omitted from the published article but that had not altered the overall results. The other retractions concerned either the omission of authors' names, incorrect listings of authors' titles, or typographical errors.

Journals may be used to publicize suspicions of misconduct as well as to correct the record. When third parties become involved in such efforts, the goals can become confused. Recent attempts by universities to "disassociate" themselves from questionable articles written by current or former faculty have raised a host of new legal questions, primarily over whether a university has the authority to insist that a journal withdraw an article. In an episode involving the conduct of Vijay Soman at Yale University, the chief of his laboratory requested the retraction of twelve papers, eight of which he had coauthored with Soman.[20] Although accusations about those papers were first made public in February 1979, the retractions did not begin to appear in print until May 1980. Even then, the retractions were made formally not by the senior scientist but by Yale University as the sponsor of the questioned research and as Soman's employer. Several journals have since indicated that they would now hesitate to honor such requests by third parties, and no regulation or law forces them to do so. The obligation to cooperate remains, in most cases, moral not legal.

Where an acknowledgment or retraction appears can also be problematic. The International Committee of Medical Journal Editors recommends that retractions appear "in a prominent section of the journal" and "be listed on the contents page."[21] Where, however, does one retract the text of books, monographs, or conference talks? Here, too, the primary journals in the field serve as notice boards. When British scientist M. J. Purves admitted to falsifying data reported in a proceedings volume, he chose to acknowledge the charges in a letter to *Nature*: "Sir—I very much regret to have to report that data published in the proceedings of the 28th International Congress of Physiological Sciences . . . are false. I must also emphasize that none of my colleagues was involved in the preparation of this paper and the responsibility was mine alone."[22]

Scientists innocently caught in these cases have often found it necessary years later to attempt to re-correct the scientific record; Efraim Racker's "update" letters to *Science* provide examples of this practice.[23] Racker wrote letters in 1981 and in 1983, in which he explained what part of the original published work was correct, how the fraud happened, and whether such incidents could be prevented.[24] He meticulously detailed his various attempts to duplicate the questioned research and stated: "I have accepted full responsibility for the confusion created by our publications and retracted the work as soon as we realized some of the data were questionable." The tone of the second letter is sad and defensive, as he explains that "the pressure [on Spector] seemed to come from inside, not from outside. I repeatedly urged him to slow down and not to work so hard."[25] By 1989, Racker felt sufficiently distant from the case to discuss it again—this time, in an article for *Nature* which emphasizes that no single explanation ever suffices when teachers, advisors, and colleagues wonder "why?".[26]

Retraction letters occasionally address questions about an entire body of data, not just a single article. For example, the letter published by Claudio Milanese and his coauthors retracted an article previously published in *Science* ("We write this letter to inform the *Science* readership about information pertinent to a Research Article . . .") but also mentioned that they are withdrawing an article from the *Journal of Experimental Medicine*.[27] The coauthors described data that are "not reproducible" and "incorrect" and apologize "to the scientific community and trust[ed] that certain misinformation presented in that article can be rectified by publication of this retraction letter."[28]

Who Does It?

In many cases, coauthors accept joint responsibility for the retraction. Who (if anyone) should accept the responsibility, moral and legal, however, for initiating retraction when not all coauthors agree? The implementation of federal regulations and the participation of attorneys at all stages of investigations can introduce concerns about the legal implications of responsibility for retraction. Accepting responsibility begins to carry financial and legal as well as moral implications. And there is an increasingly strong tendency to delay any notification whatsoever until a case is settled.

At several universities, investigating committees, impatient with the federal system's sluggishness or resistance, have communicated directly with editors of journals in which questioned articles were published. In a case at the University of Cincinnati, for example, the investigating committee recommended that Charles J. Glueck "immediately retract or issue a clarification" of his disputed article, but they also recommended that the university independently notify the editors who *might* be considering book manuscripts based on the study, should notify journals to which the accused had submitted other articles, and should notify journals in which he had published letters referring to the study.[29] The university, however, acted only on the first recommendation.

When one coauthor refuses to admit error and the other authors take the initiative of withdrawing a questionable article, the level of acrimony rises quickly. What rights do the other coauthors have in such a situation? Can a single coauthor block the publication of a retraction notice? Can a single coauthor require it? One solution adopted by coauthors facing an unwilling colleague has been to describe the pertinent events in a letter to the editor and to indicate where questions have been raised but to refrain from formally retracting the article. That solution at least alerts readers that data have been questioned. A scientist innocently caught in the Alsabti plagiarism case (chapter 1) made a similar effort to publicize Alsabti's activities. E. Frederick Wheelock, who had allowed Alsabti to work in his laboratory and whose grant application Alsabti had apparently plagiarized, wrote a letter to the *Lancet* describing what had happened.[30] Wheelock also wrote to three other journals that had published the plagiarized papers, but only the *Lancet* published his notice.[31] In another dispute involving Columbia University chemist Ronald Breslow and his graduate student coauthor, who was accused of fabrication, Breslow wrote to *Chemical & Engineering News*, noting that he could not duplicate work described in three communications published in *Journal of the American Chemical*

Society. The tone is very cool: "I apologize for any confusion this situation may have engendered, and hope that no one else has wasted time following false leads."[32] Although it was not technically a notice of retraction, the accused student regarded it as such. She continued to protest her innocence and to use the letters column of the same publication to plead her case: "Can the principal author of a paper withdraw it . . . without showing evidence contrary to the published data to the coauthor?" she asked.[33] The university committee later concluded that falsification had occurred.[34]

The University of California, San Diego committee investigating Robert A. Slutsky (chapter 4) also adopted a direct approach.[35] The faculty investigating committee reviewed 147 manuscripts, 137 of which had been published. They reported six types of misconduct: (1) fictitious accounts of tests and procedures; (2) alterations of real data (such as changes in the level significance of statistical tests or the locus and time of an experiment); (3) use of previously published data without appropriate citation; (4) listing people as coauthors without their permission or knowledge; (5) falsifying data on his resume; and (6) forging coauthors signatures on copyright permission forms accompanying journal submissions.[36] The committee attempted to contact all ninety-three coauthors of the questioned papers, asking them to indicate what role, if any, they had taken in either the research or writing.

The committee eventually classified the 147 manuscripts as either "valid" (79 papers), "questionable" (55), or "fraudulent" (13).[37] By coincidence, these categories resemble the three levels being assigned by the Rembrandt Research Project (R.R.P.), an effort by a group of Dutch art scholars to determine the authenticity of every work attributed to the artist Rembrandt. The R.R.P. applies evaluation criteria that range from technical aspects such as the counts of canvas fibers to aesthetic and historical data, in classifying Rembrandt's paintings into three categories, either A, B, or C: "A's are those paintings which are undoubtedly by Rembrandt; B's those on which no absolute decision has been reached; and C's those which the team concludes are not by the Master and are therefore de-attributed, dis-attributed, demoted, or rejected."[38] Predictably, the middle category ("questionable" for the UCSD committee and "B" for the R.R.P.) provokes the most controversy; in the Rembrandt project, many museums and private collectors whose works have been classified as B's or C's have simply refused to accept the R.R.P.'s authority to judge authenticity. In the Slutsky case, the middle category included articles for which there was no coauthor to verify accuracy or for which the original data were lost, but the classification's ambiguity left the validity of the article in limbo.

Notified of the committee's findings, Slutsky also refused to acknowledge fraud but did withdraw fifteen of his publications.

Had the UCSD episode ended with the committee's report to the university, then the case would have resembled many others described in this book. However, this particular committee decided to proceed one step further and to notify the journals that had published Slutsky's articles, regardless of the committee's classification of the work. Letters were sent to over thirty journals. Each coauthor was notified of the committee's classification of the paper, sent a copy of the committee's statement to the journals, and, as the committee stated, was "given an opportunity to correspond directly with the journals or to include a note with the letter from the investigating committee to the editor."[39] Paul Friedman, chairman of the committee, later described how the journals had responded:

> At least one expressed outrage at the poor quality of the research supervision at UCSD . . . that would have allowed such extensive fraud to exist and go undetected for so long. Some expressed complete willingness to publish the statements we furnished. . . . Others edited out the statement that we believed the author's previous withdrawal of certain papers implied an acknowledgment of fraud, whether or not the committee had classified the paper as fraudulent or questionable. Some thought it necessary to publish a statement about valid papers.[40]

Friedman has also reported that one journal informed the committee of its documented policy that "papers could only be retracted by a co-author, not by a third party or institution." About half of the journals to which the committee wrote did not acknowledge receipt of their letter. Other journals stated that they would have their attorneys review the committee's request before making any decision on publication. The committee's own negative reaction to this mixed response exemplifies a widespread misunderstanding among the academic science community about the roles and responsibilities of journals in misconduct cases. A journal must consider many factors in deciding how to respond to such a request, such as opposition from an accused coauthor, and the adjudication of interests that may conflict with the agenda of an ethics committee will not be easy.

The UCSD committee attempted to shift the burden of responsibility for correcting the scientific record away from the university and onto the journals. Moreover, it did so with the apparent expectation that the journals and the university had identical interests. Although both publishers and academics despise deception and want to preserve the integrity of the scientific literature, as one editor commented later, a journal's staff confronts terrible problems when it must decide what to do about an article labeled as neither clearly valid nor clearly fraudulent and when the author (or all coauthors) are not initiating the request for action.[41] Ambiguity of classification determines the speed and extent of response. Eighteen months after the committee had sent its letters, only some of the journals had retracted or even publicly clarified the status of the Slutsky articles.[42]

How seriously a retraction is treated is also influenced by who initiates it—the author, the journal, or some third party.[43] When an author initiates a retraction, most readers probably perceive the communication as a routine part of science; the author may be regarded as "passingly careless" but not necessarily as maliciously deceptive.[44] Readers may also tend to be relatively forgiving if an author admits culpability and confronts an error directly. When an article is withdrawn by someone else, however, there is an inescapable aura of either coercion or condemnation which stimulates "an atmosphere of distrust and suspicion."[45] Yet this is exactly what is now often being done as university committees, frustrated that a false record might stand and unwilling to accept dismissal or disbarment as sufficient punishment, seek to force journals to publish the committee's notices or retractions.

The Slutsky case represents the most elaborate effort to date of a university's organized attempt at retraction, and it illustrates the gulf between what journals can and will do and what academics expect and desire them to do. When *Circulation*, a journal sponsored by the American Heart Association, received the UCSD letter, its editor responded that journal policy required notification by an author or coauthor before a published paper could be retracted.[46] Editor Burton E. Sobel reported that the association's policy sought to protect authors "from institutions that could arbitrarily initiate retractions, thereby defaming authors and violating their due process."[47] That policy also reflected the journal's own perceptions of its hierarchy of responsibilities: its first obligation, in such a situation, was to the author and then to its own legal protection, lest it be sued for libel. *Circulation* did not accept that it had a primary obligation to another institution. The association's general counsel, responding to continued criticism from UCSD officials, noted that "a journal must balance the interests of an individual scientist against the desire to correct any erroneous information."[48] A similar situation existed at *Radiology*, from which Slutsky, through his attorney, did withdraw four articles but from which he refused to withdraw another four that the UCSD committee had classified as "questionable" or "fraudulent." Because a coauthor on one of the latter articles insisted that there was no problem with the original work, the journal's attorney recommended no further action.[49]

There is no uniformity to these practices. The International Committee of Medical Journal Editors, for example, recommends that all journals "print retractions of questioned or fraudulent research even if the lead author or co-author have not submitted or approved such statements."[50] Nevertheless, many journals continue to refrain from active cooperation, perhaps out of fear of lawsuit or because "too many" retractions might damage a journal's reputation for quality and reliability.

When an author initiates or cooperates in a retraction, the form of notice can be a statement or a letter to the editor. Usually the journal simply acts as a conduit, publishing notices without comment. In the case of three articles by John Darsee, *Annals of Internal Medicine* took a more active approach. The journal published letters from the coauthors involved asking that the articles be retracted, along with a response from Darsee; and then the editors added their own scathing editorial, "The Responsibilities of Coauthorship," to the same issue.[51]

The wide variations in responses indicate the magnitude of unresolved questions about how journals can and should balance the rights of the accused, the responsibilities and rights of coauthors, and the responsibility to notify readers that data are being questioned. Ultimately, the peaceful resolution of any case may come down to whether, when, and where a notice of some type of official judgment can appear. The process will inevitably contain elements of frustration for all parties, however, because no article, once published and mailed to subscribers, can ever really be brought back.

Technical Problems

If only the original monographs or journal articles were involved in the retraction process, the technical and legal problems would be difficult enough; but research communication stretches also throughout a vast system that indexes, abstracts, references, and disseminates these primary materials, and that creates "shadow" systems of secondary communications based on them. In addition, electronic information networks are now linking some libraries and research institutions to provide full-text versions of technical journals, as well as computerized abstract data bases, so that a journal article can now have both a print and an electronic "existence" independent of one another.

Bibliographic services include lists of every substantive publication in every field or subfield of the sciences and social sciences. Within weeks of publication, the table of contents of a journal issue may appear in *Current Contents*, abstracts or summaries will appear in bibliographic services such as those administered by the Library of Congress or the National Library of Medicine, and other information about the articles and their authors will flow through on-line information networks to libraries and individual users. Once an article is cited in print, its existence becomes part of the citation indexing system.

If any subsequent notice about that article or book does not somehow follow it back through these systems, unsuspecting readers and system users will possess only a partial truth. The problem of correction is enormous, and the secondary services do not agree on any best response. Should they uniformly attempt to remove "tainted" articles from data banks, should they tag them with some "warning label" (and, if so, when and how), or should they let the record stand in the sequence published? *Chemical Abstracts*, an indexing service owned by the American Chemical Society, took a straightforward approach when it was still primarily a print-oriented service: it simply republished a corrected abstract and then eliminated the questionable abstract from future volumes, "assigning it to oblivion" by even killing the index entry.[52] Now, as a computerized service that includes millions of abstracts and sells data tapes to thousands of vendors, the organization cannot so easily "eliminate" entries. Unlike Dr. Beringer, a data service cannot buy back every tape or diskette, only correct the current version.

Some type of consistent cross-referencing seems the most practical solution for the short term. Since 1984, one of the largest medical indexing services, the National Library of Medicine (NLM), has been linking retraction notices to the original article in MEDLINE, its electronic data base, and doing so with increasing thoroughness.[53] When MEDLINE users retrieve a reference to any article that was later retracted or corrected, the system automatically notes the subsequent action. The published version, *Index Medicus*, also includes retractions and, in recent years, many other abstracting and indexing services have followed suit. When the NLM decided to implement this tracking (rather than removing all references to a retracted article), the decision was based primarily on philosophical grounds. As NLM administrator Robin Shivers explained, "We decided not to remove cites from MEDLINE because the library didn't want to be seen in the role of censor."[54] The service chose to describe the history of an article's reputation (that is, how it was initially accepted as authentic and then later questioned or discredited) rather than to perpetuate a false history or, in effect, to rewrite the history of science by eliminating references to discredited articles.

Indexing services are hampered in their attempts to link the original article and any subsequent notice, first, by the extraordinary volume of material indexed, and second, by the inconsistency of journal retraction policies. "Correction or retraction" was not, as late as 1986, a criterion for selection of an entry for *Index Medicus*. The service included some letters to the editor but no correction notices, a policy that has since changed.

Indexers also face a jumble of terms and practices used to describe problems with texts and authors; some journals use the key word "retraction," whereas others still prefer "withdrawn." Uniformity of terms, use of the same key word in both original article and retraction notice, and listing the notice in a journal's table of contents will facilitate cross-referencing, but the users of data bases and similar services must still be on their guard, conscientious in their searches for information on an article's history and reputation. And, in the face of pressure to cleanse the literature, indexing and abstracting services will increasingly face the dilemma of how proactive they should become. Should they take action, for example, when suspicions have been raised about research (and perhaps even questioned in congressional hearings and government investigations) but no retraction, correction, or other notice has been published?

PUBLICITY: THE ROUTE OF LAST RESORT

Verbatim plagiarism may be easier to detect than forgery, but it presents some of the thorniest issues for resolution in that theft of ideas cannot really be undone, any more than a mugger can undo an assault. One cannot return stolen credit, or restore the priority of first publication. Restitution for the crime of plagiarism must follow some other course.

If the violation involves fabrication or forgery of data, and the accused repeats the experiment, the incorrect or deceptive publication has nevertheless been disseminated. Publishers and editors therefore occasionally resort to "publicity" to correct the record, just as employed in the Pro, Blackman, and Alsabti cases (chapter 1). In the face of resistance or when one cannot prosecute the wrongdoer, publicity becomes an attractive alternative. Although Harvard professor Shervert Frazier was later stripped of some of his academic appointments, the initial *Boston Globe* reports on the case appeared to assume that he would not receive adequate punishment because of his status in the university. In a box accompanying the news story, the paper therefore provided readers with "a comparison of excerpts," essentially publishing the plagiarism; and it took the same approach in 1991 when a Boston University dean was accused of plagiarism.[55] Publicity requires extraordinary sensitivity, however, and should normally be the last resort. Those who seek to publicize an alleged wrong must not only check the facts but search their own motives in acting as judge, jury, and executioner.

Another symbolic action may be taken by the research institution under whose auspices a fraudulent project was conducted and under whose affiliation the work may have been published. MIT's *Policies and Procedures*, for example, states that in addition to invoking disciplinary actions such as termination of employment, "the Provost has the authority to mitigate the effects of the fraud by withdrawing MIT's name and sponsorship from pending abstracts and papers and by notifying persons known to have relied on any work affected by the fraud."[56] Such a policy lends authority to an institution's use of the term "sponsored" research, which is usually reserved for work supported by government or private grant to the university. The policy acknowledges that work published under the name of a well-respected institution should not just be assumed to be but indeed *must be* work of the highest scholarly integrity. Because journals and books usually list an author's affiliation, the practice of withdrawing affiliation offers an interesting alternative for publicizing and punishing certain forms of misconduct.

Not all participants in misconduct investigations agree on the importance of publicizing their results—usually because they believe that secrecy is necessary to preserve the personal privacy of accused, accuser, and those who provided evidence.[57] The Office of Scientific Integrity Review (OSIR), part of the Department of Health and Human Services (HHS), initially denied Freedom of Information Act requests in 1989 and 1990 for detailed information on recent cases, arguing that all records should be kept closed; but similar requests *were* filled in 1991. When the director of OSIR first announced the policy (at a scientific meeting in 1990), the decision to deny requests was greeted with disapproval. Critics pointed out that the policy was inconsistent with how records of investigations of other types of professional misconduct, such as that of physicians, are handled. The OSIR official acknowledged that describing infractions only in general terms could tend "to exaggerate minor violations and make the problem appear more serious than it actually is," but he argued that disclosure harms all those who undergo an investigation, regardless of the seriousness of the violation.[58] One scientist who defended the policy argued that secrecy allows accused researchers to continue with their careers; to him, "disclosure constitutes an additional punishment."[59] Although forgiveness represents an important and honorable goal, Patricia Woolf agreed, those scientists who "betray the public trust" should, in her view, be considered to have forfeited any right to privacy regarding the nature of their violations and any resulting punishment.[60] This issue, too, promises to attract intense debate in the future.

Some journals that received notification from the UCSD committee investigating Robert Slutsky "expressed concern that they would become as liable as the University if they published statements impugning an individual's work."[61] In the Stephen Breuning case, even after his federal indictment, some of his articles were not withdrawn because the journals in question were fearful of being sued.[62] *The Scientist* reported that only a few editors who had been sent the government report on that case and only a few coauthors of the questionable articles "have considered, or attempted to publish, retractions."[63] More than fifty articles remained in dispute for months, because Breuning refused to discuss the matter publicly and many of the journals would not retract a coauthored article unless all authors formally approved of retraction. Marcia Angell of the *New England Journal of Medicine* commented at the time that fear of lawsuits should not "interfere" with efforts to correct the record, but most editors seemed to acknowledge that it did. Some editors argued that if an author would not cooperate, they had no choice but to let an article stand. The following spring, the International Committee of Medical Journal Editors endorsed a policy declaring that journals "must print a retraction" when notified and a spokesman for the committee dismissed fear of a lawsuit as irrelevant. Even now, however, that view does not represent the prevalent attitude.[64] In a 1988 article, two well-known attorneys argued that the "fear of liability" reportedly discouraging many editors from publishing retractions is "greatly exaggerated" because the "'fair reporting' privilege protects publication of facts that are of legitimate interest to those who would be expected to read them."[65]

FEDERAL REGULATIONS AND PROSECUTION

At present, federal regulations apply to U.S. research institutions such as universities and colleges, but not apparently to private-sector journals or their publishers, unless the latter have received federal grants for specific projects that are being questioned. National Science Foundation (NSF) regulations adopted in 1987 [45 *Code of Federal Regulations* Part 689], for example, required all institutions that receive NSF funds to establish policies and procedures for handling misconduct allegations, to complete inquiries promptly, and to inform the NSF if that inquiry determined substantive possibility that misconduct had occurred involving an NSF proposal or award.[66] Because the NSF definition of misconduct includes plagiarism, it seems likely that its applicability to projects involving journals (e.g., funding for special issues) or books (e.g., large editing projects or publication subsidy) will sooner or later be tested. Regulations issued by the Public Health Service in the Department of Health and Human Services in 1989 [42 *Code of Federal Regulations* 50.101–102] include any "entity" that "applies for a research, research training, or research-related grant or cooperative agreement under the Public Health Service Act" and also include "plagiarism" and other prohibited conduct during the "reporting" of research. As Barbara J. Culliton pointed out in a news analysis of the changing legal landscape for fraud investigations, there are two "fundamentally different" ways to approach the issue—as an internal matter similar to peer review or as a case potentially "headed for court."[67] In the latter instance, concern about due process for the accused and other procedural safeguards propel an investigation early on into formality and the tendency to involve participants, such as attorneys, who are not part of the institution and who are often not scientists.

It remains to be seen whether or not federal regulations designed for university-based research will be more broadly applied to include the editorial review and formal publishing process, but there are indications that such extension could occur. In 1988, congressional legislation was introduced to designate research fraud as a "white-collar crime" subject to federal jurisdiction. Although this particular legislative proposal failed, it raised important questions that the publishing industry has yet to confront directly. A criminal investigation could challenge the anonymity of the review process as well as test the confidentiality of relationships between editors and referees, or editors and authors. Could an editor or publisher, for example, be considered criminally negligent for inadvertently publishing fake data or failing to investigate a referee's suspicions? These hypothetical questions took on other dimensions later that same year when another congressman introduced a legislative provision to require the Secretary of the Department of Health and Human Services to "develop guidelines for use by scientific journals to protect against publication of manuscripts with respect to which there has been scientific misconduct," and to punish noncompliance by prohibiting the National Library of Medicine from identifying in its referencing systems "any journal that does not subscribe to and apply the guidelines."[68] Even though that legislation also failed, the provision (slightly rewritten) surfaced again in 1991 as an amendment to the NIH authorization legislation. These proposals signal a change in political attitudes that includes the communication of science in the political debate. The mere *possibility* of federal restrictions on communication raises significant issues of publishing rights and freedoms which extend far beyond the world of R&D.

Cases that appear to have been thoroughly investigated can still result in ambiguous outcomes. Uncertainty and ambiguity create difficult challenges for journals. In the David Bridges case, the NIH, acting on the recommendations of its own committee, barred him from serving as an NIH proposal peer reviewer for ten years and stripped him of current grant awards.[69] Although the actual arena of misconduct was the journal *Science*, apparently neither *Science* nor any of its representatives played a direct role in the NIH investigation, however.[70] In the end, *Science* was left to "clean up" the record on its own.

In other cases resolved by HHS investigation during 1989, eight of eleven scientists investigated were prohibited from serving on agency advisory panels for periods that range from three to five years. In five of the cases, the agency's ethics office also recommended that the accused scientists be barred from receiving any federal research funds for penalty periods that ranged from three to five years, and employers of some were asked to conduct "special reviews," scrutinizing the scientist's work with extra care for a designated period.[71] A later report by the HHS Office of Scientific Integrity Review, covering twenty-one investigations completed in the period March 1989 through December 1990, stated that misconduct had been found in fifteen cases and that, as a result, twelve scientists had either voluntarily resigned their positions or been dismissed. Scientists implicated in federal investigations could be asked but probably not *forced* to withdraw an article or publish a notice. At present, no independent or private-sector journal could be compelled to publish a notice of correction or retraction, although many would certainly cooperate.

Until the federal prosecution of Stephen Breuning "no previous case of alleged scientific fraud is known to have resulted in criminal prosecution."[72] Now that the Breuning case and subsequent ones have entered the criminal justice system, there seems little doubt that scientists who lie about their research in reports or proposals to the federal government may well be investigated by grand juries and may be prosecuted.[73] This circumstance has a number of implications for journals and publishers who may be involved, directly or indirectly, in future criminal investigations.

Although allegations in the Breuning case were handled informally at first, the entrance of the Justice Department altered the investigation's character as well as its political outcome. Breuning was indicted by the federal government under the False Claims Act, "which prohibits the submission of false information to the government for the purpose of obtaining money," that is, he was accused of taking money but failing to do the research.[74] The law carries a penalty of $10,000, up to ten years in prison, or both. He pled guilty and was sentenced to mandatory public service, in addition to the fine.

Another route for federal action may be specific legislation to make scientific fraud a "white-collar" crime. Commenting on such proposals, chemist Irving Warshawsky argued that although the intent appeared laudable, the law would probably not influence the behavior of young scientists or scientists-in-training; the proposed regulations were simply "too narrow in . . . focus and too punitive in . . . scope" to influence the educational process.[75] Prosecution under the act would have required the testimony of expert witnesses but scientists are notoriously reluctant to "have their lives and careers disrupted," he emphasized. Despite such objections, there are indications that similar legislation will be proposed (and may be passed) in the future. There is so little consensus on what constitutes wrongdoing, and so little adequate data on how much fraud exists, where it occurs, and so forth, that it will be difficult to construct legislation applicable to all fields and all publishing situations and that makes discussion of these issues now all the more important.[76]

ALTERNATIVE PUNISHMENTS

In the 1970s, when reports of scientific fraud received far less press attention, the employers of accused scientists could act protectively and soften the dismissal process even in the most blatant cases. When William Summerlin admitted to falsifying experiments, Sloan-Kettering responded by immediately suspending him (with pay) and setting up an investigating committee. When the committee later found Summerlin guilty of "misrepresentation," he was placed on terminal one-year leave at full pay, a penalty criticized as far too lenient.[77] Nevertheless, these same attitudes were still apparent among some in the scientific community years later. In the Shervert Frazier case, many of Frazier's colleagues charged that forcing Frazier to resign from even one of his university posts was too harsh; the penalty did not "fit the crime."[78] This lingering inconsistency in response results in continued inequity in how researchers are treated.

To some, it is disturbing that many of the biomedical scientists accused or convicted of fraud have left research but then take up the full-time practice of medicine because they have an M.D. in addition to their Ph.D. The role of physician should be one in which trust and high ethical standards are important, one writer notes; allowing those accused of research fraud to continue to practice medicine sends a message to potential physicians and patients alike that ethics in medicine "are of minimal importance."[79] The Commonwealth of Massachusetts, perhaps in reaction to such criticism, fought from 1984 to 1989 to revoke Alsabti's medical license, eventually succeeding after multiple court appeals by Alsabti.[80]

Even when committees have found error but no willful misrepresentation, they have sometimes still felt compelled to mete out punishment. In a Stanford University case, the accused received a letter of admonishment, was "advised" to consult with statisticians on any future work, and was "asked to submit an appropriate retraction" of the article questioned, and to send a copy of that retraction to the NIH study section that reviewed his federal grant.[81] In effect, he was asked to administer his own punishment. Neither the study section, the university, nor the accused reportedly informed the NIH office responsible for investigating misconduct about these punishments at the time the action was taken.

Perhaps because of such lapses, it has been suggested that journals exchange informal "blacklists" of authors known to have submitted or published fraudulent papers.[82] Although well-intentioned, that suggestion is full of potential for mistake or malicious action. Nevertheless, editors and publishers must begin to consider some type of uniform policies for handling information about alleged and proven fraud. And these policies should specify how authors, once convicted, will be treated in subsequent review, whether journals will share that information (and, if so, with whom), and whether an earlier substantiated charge of fraud will, in a spirit of rehabilitation and forgiveness, be considered irrelevant or must be factored into every subsequent review.[83]

WHO PAYS?

The human costs of fraud are well known. Lives disrupted. Careers destroyed. Personal and professional relationships torn apart. Other factors that surface toward the end of a case involve the associated economic costs. When a university researcher is convicted today of having misused federal research funds, federal auditors will probably require the university to refund the money. The University of Pittsburgh returned $163,000 to the NIMH following investigations of alleged fraud by Stephen Breuning, for example.[84] A more difficult question for the future, however, will be who will pay for the cost of investigations and for any subsequent efforts to correct the scientific record.

Some initial costs of fraud occur in the editorial office, when fear of increased misconduct causes staff to perform additional checks on all manuscripts. The cost of verifying the identity of all authors listed on a manuscript, much less authenticating their actual participation in the research and writing, far exceeds the financial and staff resources of most journals;

but more rigorous inspection of articles at the editorial level, before publication, could obviously save resources, effort, and time for readers who rely innocently on data later retracted or proven to be false.

Should the increased costs of screening and cross-checking be borne by the individual journal and its readers alone via higher subscription costs? Should costs be passed on to authors through imposition of surcharges? Or should monitoring costs be subsidized by foundations or government? Clearly, better data predicting what the additional costs are likely to be and where they are likely to be incurred will help in making these decisions.

The other major unresolved questions, such as who should bear the burden of proof when a misconduct accusation is made or who should take responsibility for action after a misconduct investigation is completed, represent assignments of responsibility infrequently discussed yet frequently interpreted differently by the scientific community and the world of publishing. At present, these potentially conflicting interests are renegotiated in each case. Closer attention to the issue by both publishers and scientists will be needed before we can even begin to locate common ground and develop workable policies.

9 On the Horizon

All this seeking for the truth of the past should be in abeyance until
we have reached some conclusions about the truth of the present.

—Angus Wilson[1]

Early in Angus Wilson's *Anglo-Saxon Attitudes*, the protagonist iden-
tifies a dilemma that will haunt him throughout the novel: what are the
links, he asks, between the "truth of the past" as it has always been
interpreted and the "truth of the present," the lies that perpetuated that
interpretation. Just such discrepancy between perceptions of "truth past"
and "truth present" pervades the discussions about scientific research fraud
in the United States today.

To many observers, the emergence of this controversy in a country
where scientists enjoy generous funding and high social status hints at a
cultural rather than a moral explanation, one related to questioning of
scientists' long-standing role as truth-seekers and truth-tellers rather than
to deficiencies of character. When scientists sweat under the hot lights of
congressional and media scrutiny, they now seem not "special" and heroic
but thoroughly human and therefore prone to the same ethical weaknesses
afflicting contemporary sports figures, religious leaders, politicians, and
bankers.

The apparent inability of many universities and scientific laboratories
to resolve investigations swiftly or equitably has added to the perception
of weakness, as has also the sluggishness of development of basic guidelines
for appropriate research conduct. Moreover, the squabbling over definitions is likely to continue—not because of the magnitude of the stakes but
because the stakeholders view the issue from such markedly different
perspectives. Scientists regard it as a science matter, politicians as politics,
agency officials as regulation, and publishers as their own private business.
Add one stubborn whistleblower who has the bit in his teeth and the
courage to walk through fire and a case may rage for months, even years.
It is symptomatic of the mismatched perceptions that, a decade after the

195

first congressional hearings on research fraud, one can participate in meetings on the subject in which reasonable, well-intentioned scientists still challenge the basic rationale for government involvement in investigations of federally funded research.

Formal definitions of expected conduct inevitably place limits on behavior. Science's past accomplishments, however, provide little guidance for determining what should be the appropriate limits to science's autonomy for the future, either in governing individual conduct or in controlling the quality of research dissemination overall. Many countries allow scientists considerable professional autonomy, justified by the argument that any restraint will inhibit all creativity, innovation, research productivity, and therefore, technological progress. Science also touches modern life in so many places—through the use of scientific evidence to justify health care programs or workplace regulation, through scientific recommendations that influence personal choice and social change, and through science's contributions to advanced military weaponry—that constraint appears counterproductive to national needs. Any proposal to limit science communication can thus ripple throughout the political economy. As this potential for political tension continues, it represents one unambiguous "truth of the present."

Those, then, are some of the broad issues. For the near future, in addition to how to formulate better definitions, there are at least five other important challenges to be considered:

1. Calculating how much unethical research conduct occurs;

2. Encouraging constructive discussion and analysis of the issues and insuring that opposing viewpoints receive fair hearing;

3. Protecting the rights of all individuals involved in cases, including accused, accusers, and investigators;

4. Anticipating how advanced communication and information technologies may influence the commission, detection, and correction of data forgery, plagiarism, or deceptive authorship; and

5. Protecting intellectual property and individual rights in an age of increased collaborative research and intensified competition.

CALCULATING THE PERCENTAGES

Is there more fraud, better detection, or just too much attention to a few unimportant cases? Determining the actual extent of inappropriate and illegal conduct remains an unanswered, and possibly unanswerable, question because assessment of appropriateness depends so much on who constructs the definition—lawyer, scientist, or publisher. Nevertheless, many scientists insist that the full extent of research misconduct must be measured precisely before any reforms can be initiated. To date, discussions about calculating the odds have progressed on two fronts, both with important implications for scientific publishing.

One suggestion centers on "data audits" or similar surveys to use editorial peer review to check for scientific fraud.[2] The proposals tend to assume that imaginary or forged data could be detected by requesting and checking the raw data said to support published claims; the extent of misconduct within a representative subset of published articles would serve as an indicator of the journal's or the field's integrity as a whole.[3]

The most well-known proposal for such an audit is one made by *JAMA*'s deputy editor, Drummond Rennie.[4] In 1988, Rennie began to suggest that a selected group of journals voluntarily cooperate to conduct "an experimental, random, confidential audit of the data behind a statistically chosen group of published scientific manuscripts."[5] According to the plan, editors at participating journals would notify all prospective authors of the possibility of an audit when articles were accepted; the editors would request that authors cooperate if they were later asked to provide data. Once the subset of target articles was chosen, professional auditors would compare the data described in the published version to the data supplied by the author. In Rennie's proposal, auditors would not "adjudicate complex and arguable scientific questions," only attest to the data's existence and, if possible, detect gross fabrication, falsification, or misrepresentation. All materials would be coded to disguise the individuals and the results for any particular paper.

Proponents of the proposal claim that data audits or retrospective surveys could demonstrate the extent of fraud once and for all. If the audits revealed substantial widespread problems, then additional federal action might be justified; if problems occurred only in certain institutions or fields, then those institutions (or perhaps the professional societies in those fields) could act to fix the problems; if audits revealed "negligible" fraud, then scientists could "feel relieved."[6]

Three important objections have been raised to these suggestions, however. How would an audit affect the climate of trust between editors and authors? How would the rights of those involved in the study be protected? How would any "detected" misconduct be treated?

Proponents believe strongly that these studies would neither threaten nor interfere with the editor-author trust relationship, but not everyone agrees. Certainly, the suggestion that authors be required to assent to such procedures as a condition of publication raises the possibility of coercion. Compliance in a government-sponsored study could also be construed as acknowledging the federal government's right to monitor research through privately owned journals, even though the misconduct of most concern (e.g., large-scale fabrication of data) would have been committed not in the journal per se but within another organization (e.g., in a university lab). Others have questioned the ethics of leaving the record uncorrected (or perpetrators unpunished) if any faked data is detected. Some emphasize that audit results could not be protected from subpoena in a court case or congressional investigation. Once it became known that a study had detected *any* instances of previously undetected fraud, government officials might understandably feel obligated to demand sufficient details to make an official investigation.

To produce statistically useful results, an audit would also have to examine several thousand papers and locate the original charts and records for all projects described.[7] An ambiguous outcome—such as one that revealed substantial "imperfect" but not necessarily fraudulent work—might cloud the reputations of the journals and publishers involved. When Rennie initially proposed his plan, he carefully stated that any experiment should be conducted under the aegis of editors, who would set a study's rules but not conduct it; but those editors would nonetheless be placed in an awkward situation if previously undetected deceptions were discovered in their own journals. A negative result could have serious financial repercussions on a journal—would the editors be the first to be notified so they could practice damage control when the results were made public?[8]

Although I question both the feasibility and wisdom of the journal-centered data audit, Rennie's proposal and similar suggestions about retrospective surveys have usefully focused attention on some crucial aspects of the measurement problem, and they emphasize the lack of reliable data about the real extent of wrongdoing. Both the National Institutes of Health and the National Science Foundation have also considered sponsorship of surveys that would use journal editors like "moral barometers," asking them to describe the types of misconduct they routinely see, to calculate the general amount they have seen, and to make a subjective judgment about whether the proportions are increasing. This type of study could then be cross-checked with a smaller but more precise and controlled one. Another measurement approach involves an agency request to a statistical sample of its grantees (perhaps stratified along size, geography, and/or prestige ranking), asking that they supply data on the number and type of incidents investigated in a particular period, and whether those incidents involved federally funded research. All these types of studies offer promising possibilities but when they involve journal publication they must be approached carefully, with sensitivity to the complex, confidential relationships involved and the potential impact of the findings.

ENCOURAGING CRITICISM AND ANALYSIS

With each passing year, the saga of the attempt by Walter Stewart and Ned Feder to publish their manuscript on the "Darsee coauthors" appears more sordid.[9] The correspondence record shows that the two biologists were frequently encouraged to press their case by people who had nothing to lose by urging them on. Editors approached by the pair became overprotective of their journals. Only John Maddox really did anything other than talk. Eventually, it took dramatic intervention by congressional committees and the press to resolve the impasse, because science, as an intellectual community, had few means to conduct productive, responsible debate on its own ethical issues.

There is clearly a need for such forums. As competition intensifies throughout the research system, disputes over published criticism of technical work appear on the increase. During the controversy over "cold fusion," for example, a lawyer representing Stanley Pons and Martin Fleischmann threatened to sue scientists who had openly criticized the chemists' work in scientific journals. The attorney argued that, although he had no desire to restrict the critics' academic freedom, he nonetheless believed it was acceptable to attempt to muzzle "inappropriate" criticism. "The academic freedom of Mr [――――] ends where the nose of my client begins," he was quoted as explaining; "academic freedom does not mean absolute right to say what you want."[10] Although the lawyer proceeded no further with his threat, the contretemps signaled a troubling new possibility that when economic stakes are raised and scientists must attract ever more private investment, the protection of one's scientific reputation from legitimate but harsh criticism may overwhelm the inhibitions, tradition, and good manners that restrained attempts at censorship in the past.

Another area of growing concern involves attempts by coauthors to correct the published record through retraction of an article or notification of the discovery of error. Patricia Morgan of the American Association of the Advancement of Science has pointed out that editors tend to see three competing purposes for corrections: to aid the progress of knowledge; to comply with federal regulations or political pressure that request timely notice of fraud misconduct; and to protect the reputations of the innocent. The conflict between these goals, when some but not all coauthors wish to retract a paper, raises several issues. Even under the best of circumstances (for example, when all coauthors agree that a problem exists and cooperate to repeat experiments or calculations), she points out, the time elapsed from first notice to an editor that a problem exists to whenever a retraction or correction actually appears can stretch for months. During that period, few journal editors notify readers that questions have been raised; does silence represent a failure of editorial responsibility, Morgan asks? She believes that fear of lawsuits may also lead some editors to hesitate to print *any* retractions, even if investigating committees or research institutions ask them to do so, an attitude she calls the editorial equivalent of "defensive medicine."[11]

Even reporting about ongoing investigations or decisions in fraud cases may now draw complaints, requiring journals to balance open and honest discussion of wrongdoing with sensitivity to the rights of all involved, both accuser and accused. As a result, major journals like *Science* reportedly ask their attorneys to review all unflattering news reports for potential libel, because truth and accuracy cannot prevent a journal from being sued or bearing the cost of defending against a lawsuit. Under U.S. law, a statement of opinion is, in theory, not defamatory if it is based on fact, especially if this factual basis is clearly spelled out, yet recent court rulings have muddied these definitions, and editors must look cautiously even at material traditionally labeled "commentary" or "opinion," such as editorials and book or software reviews.[12]

Editors in the social sciences who wish to publish legitimate scholarly analyses of fraud and misconduct face additional potential legal problems. The controversy surrounding the Stewart and Feder article emphasized the vulnerability of journals and journal editors. A number of U.S. legal rulings have further increased the concern that use of libel law could inhibit legitimate criticism and stifle publication of controversial scientific material.[13] If no official report of an investigation is made public and the accused continues to claim innocence, then journals might reasonably refrain from discussion of a case or refuse to retract an article unless all coauthors agree.

New legislation should not be necessary to protect the First Amendment rights of periodicals to report accurately, honestly, and in good faith on the results of legally constituted investigations, nor should special legislation be needed to protect scholars who conduct serious research and analysis on unethical behavior, wherever it occurs. There is sufficient evidence to indicate, however, that reaffirmation of those rights may be necessary to preserve a free *and open* system of scientific communication.

PROTECTING THE INNOCENT

How can we adequately protect the rights of individuals involved in fraud and misconduct cases, including those who are accused, those who make allegations in good faith, those who are appointed to investigating committees, and those who, because of their positions as editors or administrators, must participate in the correction or punishment of alleged wrongdoing? The actions of the University of California, San Diego (UCSD) faculty committee that investigated the allegations against Robert Slutsky, as discussed in previous chapters, demonstrate the range of conflicts that investigators may routinely experience.

This particular committee made a heroic effort to learn the truth about troubling allegations. They were, however, new to the process of investigating fraud; they were neither lawyers nor experienced criminal investigators; instead, they were scientific colleagues chosen for their knowledge of the technical subtleties of Slutsky's work. The committee "adopted the position that the nature of scientific research requires that hypotheses must be proven and observations validated." It wrote: "In the committee's judgment, the legal principle of 'innocent until proven guilty' does not apply to scientific work; the burden of proof is on the authors of a challenged scientific publication to establish the validity of their results."[14]

Presumption of guilt flies in the face of the American system of justice, however, in which the burden of proof is traditionally placed on the accuser or the state, not on the accused. One can empathize, of course, with the frustration and anger that must have been felt by the academics who encountered stonewalling and had no means to force cooperation. One might even accept that the model of "guilty until proven innocent" adapts the approach taken by thorough and diligent scientists when they confront new evidence. Nevertheless, adjudicating the guilt or innocence of a human being, especially given the negative consequences of even having been the subject of an ethics investigation, demands an exceedingly sensitive standard. The UCSD committee's good intentions are not in question. They had set forth into unfamiliar territory. Their experience, however, provides a guideline for developing alternative approaches in the future. The professional reputation of any person publicly accused of serious fraud or ethical misconduct will suffer some harm, no matter what the eventual outcome. Even if the verdict is "innocent" or "not proven," life will probably not return to normal. The status quo will simply not be retrievable.

Another factor that affects the thoroughness of an investigation and that might bias its outcome is the hesitation of academic institutions to conduct any investigation whatsoever. As in C. P. Snow's *The Affair*, reluctance may stem from community opinion favorable to one party or another; if a whistleblower appears "motivated by essentially personal considerations such as redress or . . . grievances, animosity, or by . . . unstable personalities," then that perception affects the institution's reaction.[15] Harold P.

Green, a law professor who has defended many whistleblowers, emphasizes that, even when truth is on her side, an accuser runs the risk of a ruined career, the "hostility of colleagues, academic censure, dismissal, professional ostracism, and lawsuits."[16] In the opinion of Green and other attorneys, innocent whistleblowers may be far more "exposed and vulnerable" than the people they accuse of wrongdoing, especially if the accused is a high-ranking member of a university's faculty or a senior and influential person in the field.[17]

Initially protecting a complainant's anonymity may help, but that is rarely possible for very long. Moreover, in the American justice system, an accused is entitled to know and confront his or her accusers. Now that federal regulations require formal investigations into every allegation involving a government-funded research project, sensitivity to the legal rights of participants must keep pace. Toward that end, a variety of proposals are being discussed, including protecting responsible whistleblowers against retaliation and assuring that all parties to a dispute have access to counsel.

ANTICIPATING THE IMPACT OF NEW TECHNOLOGIES

Some writers cling to their old manual typewriters; others couldn't imagine life without computers, E-mail, or fax; even the traditionalists, however, tend to see few negative aspects to the technologies now available for research communication. Unfortunately, the same technology that makes life easier for the honest researcher may also assist the dishonest one. Rapid and easy dissemination will facilitate plagiarism, the fabrication of data (including data bases), and attempts to obscure authorship or authenticity. In addition, the increased use of computers to mechanize team-written reports may influence how teams assign and accept responsibility for the integrity and accuracy of the entire text.

In some instances, conflicting practices and expectations are the result of extraordinary changes in how research knowledge can already be communicated within and out of science. Because the procedures and policies that govern how manuscripts are reviewed, published, and disseminated evolved in a print-based era, the standards assume a print-based model of professional interactions in which there are a physical object (the manuscript) and temporal and physical limits on speed and ease of dissemination of unauthorized copies. In many fields, however, on-line electronic dissemination of research results already augments or even supplants print-based communication.[18]

Electronic mail and computer networking alter informal communication among researchers.[19] The technologies reshape the form as well as the speed of research communication; the roughest suppositions can be quickly assembled and printed such that they appear sophisticated *and complete*. Moreover, no print copy need ever exist. Commercial and nonprofit electronic publishing ventures range from services that store the entire texts of journals and allow users to read and print copies of particular articles at their personal work-stations, to CD-ROM storage of thousands of reference works. The billion-dollar on-line publishing industry and the establishment of new refereed "on-line" journals add other complications to dealing with the consequences of fraud or error, such as how to retrieve "bad" information or make corrections.[20]

Each change in the system brings potential new benefits but does not automatically improve quality. Pergamon Press has experimented with faxing manuscripts and reviews between the editors and referees of several of its journals; its study showed that a fax link reduced delays but did not produce more thoughtful reviewing.[21] Nevertheless, the pressure for electronification of manuscript processing and reviewing is intense in some fields. According to the conventional romantic model of a peer-reviewed journal, the editor and a group of conscientious referees carefully read every word in every manuscript, cross-checking facts or statistics about which they have questions, and perhaps even re-reading the manuscript after each revision. That stereotype, although fanciful, reflects the orderliness and conscientiousness that electronic review systems, under pressure from authors, may not be able to achieve. The stakes of first publication seem so lucrative—perceived Nobel prize or patent at one extreme, and tenure or promotion at the other—that authors in "hot" research fields are pushing top-ranked journals into competition for even faster review. *Science* has cited data for the review turnaround times of what it labeled the "hottest" scientific papers of the last thirty years.[22] The times ranged from three months to as little as four days. The most troubling statement in the *Science* report, however, was that "everyone knows that the way to get fast service at *Science* or *Cell* or *Nature* . . . is to tell the editor that a competing paper is coming out in one of the other journals."[23] One obvious "truth of the future," therefore, is the hidden influence of competition. We might well consider whether there is such a thing as *over*accelerated review, insufficiently conscientious and thorough, and apt to miss errors, inconsistencies, or worse.

Some technological innovations create new opportunities for misconduct. The use of digital scanners and the electronic manipulation of text undeniably speed the writing process, but those same technologies can make plagiarism (through the artful "reworking" of texts) terrifyingly easy and harder to detect. Storing original documents and reference works electronically will likewise affect how and whether reviewers or readers can verify a manuscript's information sources or citations. The use of a citation implies that the original document is available somewhere for inspection and validation; but if source data are electronically stored, they might also be accidentally, maliciously, or deceptively altered and erased.[24]

A new locus for forgery may be the computer- or instrument-generated photographs that serve as critical evidence in fields such as physics, chemistry, or biology. The editor of the *Journal of Histochemistry & Cytochemistry* recently articulated a widely shared fear: that because computer-generated or computer-altered illustrations have no negatives, there is little way to countercheck an image's authenticity.[25] Even museum curators of photography, whose work centers on the authentication and verification of historical and art photographs, privately express concern that advanced computer technology (e.g., video synthesizers) can produce altered images that are virtually undetectable by experts.

As the management of research changes, so does its communication, especially in terms of who does or does not participate in the process. In explaining one difficulty in detecting art forgeries, critic Nelson Goodman distinguishes between *autographic* works of art (those done in the artist's own hand) and nonautographic, or *allographic*, works (those inscribed by another person for the artist).[26] Music, for example, is generally allographic; the originality of a symphony (its authorship) are separate from its score and performance, both of which may require the intervention of experts other than the composer, such as transcribers, printers, musicians, and conductors. Similarly, an engraving, art print, or sculpture cast in metal may require the intervention of a technical advisor or craftsman. Particle physics, high-energy laser research, and astronomy, for example, all use enormous, complicated instruments and facilities that require, for their operation, dozens of skilled technicians; reports from projects in such fields are increasingly "produced" by the team rather than by any single author. As shifts toward "allographic science" continue, will the concept of an autographic work of science, as embodied in the traditional article, be eroded? The article, in its written form, is the conventional end-product of research. Will this, too, change?

Contemporary research increasingly fits a "two-stage" model, in which laboratory work is one thing and the communication of its results another. In that second stage, some authors or coauthors may have never entered the laboratory or participated in the data collection or interpretation. The problems raised by this artificial separation of data creation and data communication are not only ones of credit and responsibility (as outlined in chapter 4) but also ones of how to define ownership of intellectual property resulting from the work.

One cannot blame plagiarism on the photocopying machine or data forgery on the computer. Although a new technology may make "new" crimes (e.g., the creation of computer "viruses") or new choices possible, the motivations for immorality or irresponsibility transcend time and technical progress. Given that communication technologies offer opportunities for misuse in science, astonishingly few processes have been implemented to detect it. When the Xerox Corporation and other firms introduced fast, inexpensive photocopying in the 1960s, the new machines threatened authors and publishers because they loosened a proprietor's control over copies of a work.[27] The technology prompted revision of the copyright law. A totally electronic publishing system for science could similarly alter traditional relationships among participants, affect the types of fraud committed, and thus restructure how and whether unethical conduct can be detected, investigated, or corrected.

ESTABLISHING AND DEFINING OWNERSHIP

The interconnectedness of communications systems will create another set of new problems, just as will dynamic changes in the social organization of science (most notably the emphasis toward collaborative research). It is hard enough to control information moving into and within a single organization's computer system; electronic interconnections among research organizations and networks, nationally and internationally, enhance information sharing in extraordinary ways and enhance creative research collaborations; they also decrease control. The interconnections make it intrinsically more difficult to define and protect the boundaries of the traditional legal and economic compartments of "ownership" and "authorship" in intellectual property. In the 1990s, these changes have been layered on top of changing international conventions to recognize "moral rights" in authorship in addition to traditional legal rights.[28]

Another set of provocative questions is being raised about who owns the raw data that represent the conventional products of research.[29] Who can use that data and who has the "right" to publish it first—all researchers in the project, only the principal investigator, or only the institution sponsoring the research?[30] In 1989, a U.S. appellate court reversed an earlier court ruling regarding a claim of ownership on a scientific paper. In the case *Weissman v. Freeman*, a scientist sued her former mentor and collaborator, asking the court "to affirm her intellectual property rights" to a paper she wrote in 1985. The paper had been built on the collaborative research of both scientists, but the younger scientist argued that she was sole author of this particular version and had distributed it to students in her review course. She charged that the senior scientist had removed her name from the paper's title sheet, substituted his own, and then distributed the paper as if he were the sole author. Although the suit was for copyright infringement, the dispute between the former collaborators, as Harold Green has observed, was only peripherally about copyright but, instead, "about proper standards for ethical conduct in science" and about allocating credit.[31] The case exemplifies how the legal aspects of a case, along which an official investigation must proceed, may be quite different from the real bone of contention among the disputants.

The ambiguity of conventional intellectual property policy on research processes and products may further complicate these controversies. Research *ideas* are most often expressed in a patent. Research *data*, however, have been considered to be something else. Sometimes the data form the evidence and proof for the ideas, but uninterpreted data have been traditionally considered to be worthless, hence the term "raw" data. Research data may be contained ("written up") in laboratory notebooks, data bases, or other project records and it includes not just numbers but also descriptions, explanations, and illustrations. Although many researchers assume that they "own" the notebooks, sketches, photographs, forms, and so forth that they produce during their research, that assumption is now being questioned in several political and legal forums.[32]

Direct legal challenges to individual ownership of the raw products of research come from two directions. First, who has the right to publish data resulting from publicly funded research—any current or past collaborator in the research, or just the principal investigator? What interests do the sponsoring institution or the funding agency have? Second, who can (or should) control access to data from a publicly funded project? Both questions have been raised in the context of government investigations of scientific misconduct where investigators have demanded access to laboratory records. Requests from congressional subcommittees and federal agencies have outraged scientists who believe that a researcher, as "author" of a data notebook, has "copyright in it" and can legitimately control access to it. Other observers, including members of Congress, have replied that research notebooks and similar records purchased and compiled with the help of government funding should not be considered anyone's private property. Margot O'Toole bluntly articulated her disagreement with scientists who espouse a narrow view of ownership: "Our notebooks are paid for by the government. . . . The NIH treats scientific data for publicly supported published work as one would a personal diary. I think the policies that emanate from this attitude should be changed."[33]

The grant and contract agreements used by most federal agencies clearly define the government's right to have access to records generated under its sponsorship, but such provisions have usually been enforced only for financial or administrative records. To date, no judicial ruling or federal regulation has clarified whether *any* government sponsorship of a large, ongoing research project would affect *all* data and records collected by a project. As collaborative projects involving several institutions, several sponsors, even several nations, multiply and as more and more projects use electronic "notebooks" to record and share their data, it will become ever harder to apply print-era criteria in determining authorship and ownership and in clarifying who should take responsibility for alleged wrongdoing (and hence who should be investigated).

Computer networking capabilities raise other issues related to defining "authorship" and determining "originality." It is now common practice in some research groups for all reports to be revised only on computer. As participants move on and off projects, or as their participation in the writing stage changes, the boundaries of creation shift continually; a single paragraph in a technical report may have been revised by dozens of people over the course of months. Some parts of a text, such as conventional "boilerplate" descriptions or lists, may be essentially "machine-generated." In an electronic work environment, the print-based legal concept of a distinguishable work with an identifiable author becomes quickly outmoded. And the harder it becomes to define expected conduct, the harder it becomes to assign responsibility for forgery, fabrication, or plagiarism.[34]

Another foreseeable problem will occur as discrete copyrighted works, such as photographs or journal text, are loaded wholesale into computer networks.[35] Without question, considerable advantages accrue from these systems: users can be notified of the journals currently on-line and can request that the full text of articles be sent to their individual workstations. In an "electronic library," resources can be available swiftly and concurrently to many different users; archive and information storage costs are reduced. There are also a number of potential problems, however. In a library, control over who uses what is most possible at the boundaries, as the information moves from storage to user, or user to user, out of one zone of control and into another.[36] Under the old-style system, uninhibited (and, more important, *unmonitored*) browsing in the stacks or card catalog was possible; in the electronic library, the calculation of royalty or use payments (or even routine data management) will require a record of the specific activities of each user. Even a benign system would involve some invasion of privacy if it records what information is used, by whom, when, and how often.

Illegal copying still attracts the most attention, probably because it threatens an immediate loss of royalty. Copyright extends only to a work's *expression*; so, the electronic storage and use of entire volumes of journals will create new potential for systematic violation of copyright "with all evidence of copying erased."[37] Many observers fear that the widespread lack of adequate copyright protection could remove important economic incentives for development and innovation or, at the least, justify ever higher subscription prices. And if electronic resources are, as is likely, available only to those with sufficient financial resources to purchase them and insure their maintenance, then serious questions relating to the equity of availability of scientific information must be addressed.

Some years ago, David Ladd, the U.S. Registrar of Copyrights, warned against framing new information policies in ways that treated authorship rights narrowly, because new technologies will swiftly render them obsolete.[38] A broad approach is especially important in developing technological responses to fraud and plagiarism in scientific publishing. National computer networks will soon be a reality, yet discussions of remedies for research misconduct still proceed as if science is stuck in the time of Gutenberg or Galileo. In the coming decades, the move from print to public information networks must stimulate reevaluation of what authorship and ownership of intellectual property in science means, as well as honest appraisal of how to detect intrusions on legal ownership without invading individual privacy.

FUTURE ACTIONS

Compared to other professions and occupational groups, scientists in the United States enjoy considerable personal liberty and entrepreneurial freedom. The publications and publishing organizations in science have also benefited from this independence and autonomy. Until now, much of the pressure for action has come from Congress through the National Institutes of Health and the National Science Foundation or has been taken, in response to federal pressure, by universities and colleges. Two other groups—the professional associations and the publishing industry—have equally clear interests in the issue but have not participated fully in either the debates or the implementation of preventative measures. All four sectors need to work in tandem to improve the prevention, detection, investigation, and correction of misconduct, wherever it occurs.

The publishing industry and coalitions of scientific editors from all fields need to become more actively involved in advising Congress on the implications of proposed legislation. In conjunction with the professional associations, these groups should consider setting up task forces, to draft and debate guidelines for communications conduct. As David H. Johnson, executive director of the Federation of Behavioral, Psychological, and Cognitive Sciences, notes, "whether acknowledged or not, a prime role of any scientific society is to define, disseminate and enforce the requirements for membership in the society," a role that carries with it important responsibilities to set standards for its members' conduct.[39] Each individual professional association can best judge the standards that may be special to the type and setting of its members' research; although there are certain general factors in common, we should not expect identical rules to apply to both geologists and geometricians.

Professional associations can also serve usefully as the neutral mediators between whistleblowers and institutions or federal agencies; they may also launch or sponsor investigations when the work involves independent scholars or non-federally funded research. Excellent models exist in many professional groups, especially those like psychology or engineering that include both academics and researchers employed in the private sector. The American Association of University Professors and other university groups have developed useful guidelines or have been instrumental in sponsoring conferences on these issues.

Organizations of editors and publishers, such as the Council of Biology Editors or the Society for Scholarly Publishing, have also begun to take active interest; many of them sponsor sessions on ethics at their annual meetings. Editors confront the consequences of ethical disasters on the firing line, but any widespread change in procedures will have to be a joint effort, with some coordination of effort among journals. Publishers, too, cannot stand back and pretend problems do not exist. By forcefully articulating their own expectations of conduct to the scientific community and by explaining how these expectations relate to individual scholarship, the publishers can take the lead and stimulate national and international attention to the discussion of publishing ethics, perhaps sponsoring the development of curriculum units and discussion materials for use in the science curriculum. The various associations for editors, scholarly publishers, and university presses have begun this dialogue at their recent meetings and it must be continued.

ETHICS AND ETIQUETTE

The late U.S. federal judge David L. Bazelon, in an essay on "Morality and the Criminal Law," raised several points relevant to discussion of unethical or illegal conduct in scientific research. He wrote: "Those who see the law as a moral force insist that the law should not convict unless it can condemn."[40] To convict, we must be able to determine that the accused has indeed committed "a condemnable act"; that the accused "could reasonably have been expected to have conformed his [or her] behavior to the demands of the law" because the law was clearly and publicly stated; and that "society's own conduct in relation to the [accused] entitles it to sit in condemnation . . . with respect to the act . . . committed," that is, those investigating and prosecuting the cases have the right to do so.[41] Bazelon's guidelines reinforce the importance of constructing unambiguous definitions of expected behavior and of encouraging open discussion of those expectations. They also emphasize how difficult equity and objectivity may be to achieve.

A standard argument in criminal law is that increasing the "cost" of a crime should decrease its commission, but this approach represents an "externalist" approach, in which one attempts to influence behavior from the outside. For violations of acceptable conduct in the communication of science, external pressure in the form of legal sanctions and government regulation must certainly be applied, but internalist methods, such as "ethical mentoring" and raising the level of ethical awareness overall, in laboratories and in the classroom, are equally important. The former requires wisdom in implementation; the latter will require continuous vigilance, unwavering resolve, and a willingness to state openly which actions will be excused and which will be condemned.

One type of conduct—exploiting the ideas of another—is not, for example, on its face, necessarily immoral *or* unethical; the principal considerations in that judgment must be the knowledge and consent of the original "owner" of the ideas as well as the adapter's intent to deceive. Researchers who publish do not and indeed may never know whether others ever use their ideas, but they hope so. To publish is to *share* knowledge as well as to disseminate it.

For all types of misconduct, investigators can quickly become immersed (and lost) in detail. A common source of misunderstanding in plagiarism cases is the form of proper attribution. Endless variation exists in how a source should be acknowledged, depending on the type of communication, the publisher's requirements, and the audience's needs, among other things. What will be sufficient information for a speech will rarely be sufficient for a journal article even though the purposes of citation are the same. "A borrower may acknowledge his source," Alexander Lindey observed, or "may decide to make no reference to it, [but] the ethical nature of his act or omission will depend on (a) the amount of his appropriation,

(b) whether the source is so familiar as to need no specification, and (c) the form of the acknowledgment, if there is one."[42] Or as Mallon phrases it, "to acknowledge something on the copyright page is all well and good; neglecting to put quotation marks around much of what is used is something quite different."[43]

Ethics and etiquette are always delicately balanced. In the end, no author can prevent theft of ideas or credit but every writer can avoid stealing them. Ideas, Walter Redfern points out, "are common property, therefore what is someone else's is also mine for the taking."[44] The lack of absolute protection, legal *or* social, against such theft makes all the more important the establishment of standards condemning deceptive publication and promoting integrity. The progress of science demands free communication of ideas among scientists; the social system of science relies on publication as part of the evaluation process for both people and ideas.[45] Scientists cannot work as scientists and refrain from publication; yet once an idea is put into circulation it "becomes pilferable"—"You can't fence in a story or chain it down," Lindey writes.[46] A moral climate that rejects stealing, defines it as a reprehensible act, and condemns the thief, instead of a climate that politely ignores plagiarism or fabrication or that excuses it according to the offender's status, will benefit everyone.

In an ideal situation, then, *authors* should be the ones who first address questions about the accuracy, credit, and authenticity of their data or texts. *Authors* should confront the debts they owe and pay them openly and cheerfully, not wait for editors (or, worse, the courts) to request payment.

Judge Bazelon's essay also reminds us that a society and those it condemns are locked in an intimate personal relationship. Fallible people construct rules; fallible people break them. The history of science is replete with examples of competent researchers who sincerely believed what they saw and, in consequence, prompted hundreds of articles to be published about phenomena that we now know never existed, or never existed as described. The possibility of honest misperception or of accidental error during research or communication obliges editors, journals, and investigating committees to proceed always with caution before labeling any manuscript as "fraudulent" or "plagiaristic" or before classifying any act as "unethical." Nevertheless, neither prudent caution nor misplaced loyalty should be allowed to dampen our genuine outrage at intentional, calculated attempts to steal into print.

Notes

CHAPTER 1: WHEN INTERESTS COLLIDE

1. Quoted in George Blackman, *On Bartholow and Pro's "Liberal Use" of Prize Essays; or Prize-Essaying Made Easy and Taught in a Single Lesson* (Cincinnati: Wrightson & Co., 1868), 1–2. [Pamphlet in the collection of the Library of Congress.]

2. Barbara Carton, "The Mind of a Whistle Blower," *Boston Globe* (1 April 1991): 43.

3. The first case to receive substantial general news coverage in the United States was probably that of William Summerlin, a Sloan-Kettering scientist accused in 1974 of faking a skin graft on a mouse. Joseph Hixson, *The Patchwork Mouse* (Garden City: Anchor Press/Doubleday, 1976).

4. Quoted in Alexander Lindey, *Plagiarism and Originality* (New York: Harper and Brothers, 1952), 232.

5. Remark made about Robert A. Slutsky in Eliot Marshall, "San Diego's Tough Stand on Research Fraud," *Science* 234 (31 October 1986): 534.

6. Marcel C. LaFollette, *Making Science Our Own: Public Images of Science, 1910–1955* (Chicago: University of Chicago Press, 1990).

7. Reviewer in the *Lancet*, English edition (5 June 1858): 555; as quoted in Blackman, *On Bartholow*, 1–2.

8. Blackman, *On Bartholow*.

9. Ibid.

10. Ibid.

11. Ibid., 2.

12. In the 1860s, the United States offered foreign authors little protection against such violations of their rights. Although both Britain and France had by then strengthened their laws to include "protection to all written works wherever they had been written," the Americans had yet

to follow suit. Thomas Mallon, *Stolen Words: Forays into the Origins and Ravages of Plagiarism* (New York: Ticknor and Fields, 1989), 40.

13. Blackman, *On Bartholow*, 1.

14. "An Outbreak of Piracy in the Literature," *Nature* 285 (12 June 1980): 429. See also William J. Broad, "Would-Be Academician Pirates Papers," *Science* 208 (27 June 1980): 1438–1440.

15. Allan Mazur, "Allegations of Dishonesty in Research and Their Treatment by American Universities," *Minerva* 27 (Summer–Autumn 1989): 189.

16. "An Outbreak of Piracy," 429–430; "Must Plagiarism Thrive?," *British Medical Journal* 281 (5 July 1980): 41–42.

17. Arnold S. Relman, "Lessons from the Darsee Affair," *New England Journal of Medicine* 308 (1983): 1415–1417.

18. These conclusions are based largely on my own reading of the drafts, as well as opinions shared by others who have read them.

19. For other interpretations of these events, see Daryl E. Chubin, "Allocating Credit and Blame in Science," *Science, Technology, & Human Values* 13 (Winter 1988): 53–63; and Ralph D. Davis, "New Censors in the Academy," *Science, Technology, & Human Values* 13 (Winter 1988): 64–74.

20. Daryl E. Chubin, "A Soap Opera for Science," *BioScience* 37 (April 1987): 259–261.

21. U.S. Congress, House Committee on Science and Technology, *Research and Publications Practices*, Hearings before the Task Force on Science Policy, 14 May 1986, 99th Congress, 2d Session (Washington, D.C.: U.S. Government Printing Office, 1987), 2–201.

22. Kim A. McDonald, "Chemists Whose Cold-Fusion Claims Created Furor Ignite Another Controversy at the University of Utah," *Chronicle of Higher Education* 36 (6 June 1990): A6–A7.

23. John Maddox stated later that he believed the first five drafts were "plainly libelous" and implied that he had prolonged a decision on the manuscript because of legal concerns, not in order to block publication. Quoted in Bernard Dixon, "John Maddox Offers Surprising Insights into His *Nature*," *The Scientist* 2 (13 June 1988): 14.

24. Daniel S. Greenberg, "Tale of the Fraud Study That's Too Hot to Publish," *Science & Government Report* 16 (15 March 1986): 1.

25. Walter W. Stewart, letter to author, 11 July 1985, noting that the manuscript was enclosed "for your informal consideration."

26. Floyd Abrams, "Why We Should Change the Libel Law," *New York Times Magazine* (29 September 1985): 34, 87, 90–92. This article is widely credited with increasing the pressure *for* publication of the Stewart and Feder manuscript.

27. Daniel S. Greenberg, "Part II: The Fraud Study That's Too Hot to Publish," *Science & Government Report* 16 (1 April 1986): 1.

28. Philip M. Boffey, "Major Study Points to Faulty Research at Two Universities," *New York Times* (22 April 1986): C1, C11.

29. House Committee on Science and Technology, *Research and Publications Practices.*

30. Walter W. Stewart and Ned Feder, "The Integrity of the Scientific Literature," *Nature* 325 (15 January 1987): 207–214. Also see Eugene Braunwald, "On Analysing Scientific Fraud," *Nature* 325 (15 January 1987): 215–216; and "Fraud, Libel and the Literature," *Nature* 325 (15 January 1987): 181–182. News coverage and comment on the case includes: Andy Coghlan, "Researchers Roll Back the Frontiers of Fraud," *New Scientist* 113 (22 January 1987): 23; Barbara J. Culliton, "Integrity of Research Papers Questioned," *Science* 235 (23 January 1987): 422–423; Boyce Rensberger, "Fraud, Laxity in Research Are Detailed," *Washington Post* (27 January 1987); Nicholas Wade, "Fraud and Garbage in Science," *New York Times* (29 January 1987); Sharon Begley, Mary Heger, and Shawn Doherty, "Tempests in a Test Tube," *Newsweek* (2 February 1987): 64; Barbara J. Culliton, "A Bitter Battle Over Error," *Science* 240 (24 June 1988): 1720–1723; and Barbara J. Culliton, "A Bitter Battle Over Error (II)," *Science* 241 (1 July 1988): 18–21.

31. Redfern writes: "The growing number of charges concerning plagiarism in the England of the seventeenth century, and the vigorous self-defence against them, indicated no doubt a growing sense of literary property-rights, but, perhaps more significantly, it betokened a ferment of ideas, a vital intellectual culture." Walter Redfern, *Clichés and Coinages* (London: Basil Blackwell, 1989), 66. Also see Ian Haywood, *Faking It: Art and the Politics of Forgery* (London: The Harvester Press, 1987); and Mallon, *Stolen Words.*

32. William J. Broad and Nicholas Wade, *Betrayers of the Truth: Fraud and Deceit in the Halls of Science* (New York: Simon and Schuster, 1982); Alexander Kohn, *False Prophets: Fraud and Error in Science and Medicine* (New York: Basil Blackwell, 1986); and Daryl E. Chubin, "Misconduct in Research: An Issue of Science Policy and Practice," *Minerva* 23 (Summer 1985): 175–202.

33. Ladislao Reti, "Francesco di Giorgio Martini's Treatise on Engineering and Its Plagiarists," *Technology and Culture* 4 (Summer 1963): 262.

34. Ibid., 289, 291–292.

35. Richard P. Suttmeier, "Corruption in Science: The Chinese Case," *Science, Technology, & Human Values* 10 (Winter 1985): 49.

36. Merrill McLoughlin, with Jeffrey L. Sheeler and Gordon Witkin, "A Nation of Liars," *U.S. News & World Report* 102 (23 February 1987): 54.

37. Jonathan L. Fairbanks, "The Art That Needs Another Look," *Yankee Magazine* 53 (November 1989): 84.

38. Daniel E. Koshland, Jr., "Fraud in Science," *Science* 235 (9 January 1987): 141.

39. See, for example, Donald A. B. Lindberg, "Retraction of Research Findings," *Science* 235 (13 March 1987): 1308; Raymond R. White, "Accuracy and Truth," *Science* 235 (20 March 1987): 1447; Michael R. Rosen and Brian F. Hoffman, "NIH Fraud Guidelines," *Science* 235 (27 March 1987): 1561; and Daniel E. Koshland, Jr., "Response to Letter to the Editor," *Science* 235 (20 March 1987): 1447.

40. Data on the number of cases investigated by these offices in 1989 and 1990 may be found in David L. Wheeler, "U.S. Has Barred Grants to 6 Scientists in Past 2 Years," *Chronicle of Higher Education* 37 (3 July 1991): A1, A6–A7; Office of Inspector General, *Semiannual Report to the Congress* (Washington, D.C.: National Science Foundation, 1989) [Number 1 reports on the period April 1, 1989–September 30, 1989; subsequent reports contain sequential six-month summaries].

41. I discuss this phenomenon at length in LaFollette, *Making Science Our Own*.

42. Clark Kerr, "The Academic Ethic and University Teachers: A 'Disintegrating Profession'?," *Minerva* 27 (Summer–Autumn 1989): 141.

43. Suttmeier, "Corruption in Science," 59.

44. Pamela S. Zurer, "Misconduct in Research: It May Be More Widespread than Chemists Like to Think," *Chemical & Engineering News* 65 (13 April 1987): 10–17.

45. The most complete description of the case is Hixson's *The Patchwork Mouse*, which includes the full text of the investigating committee's report and Summerlin's statement.

46. Hixson, *The Patchwork Mouse*, 220.

47. Ibid., 216.

48. Barbara J. Culliton, "The Sloan-Kettering Affair (II): An Uneasy Resolution," *Science* 184 (14 June 1974): 1157.

49. Dennis Rosen, "The Jilting of Athene," *New Scientist* 39 (5 September 1968): 499, wrote that many of those who committed fraud were "mentally unbalanced."

50. Barbara J. Culliton, "The Sloan-Kettering Affair: A Story without a Hero," *Science* 184 (10 May 1974): 644–650.

51. George Savage, *Forgeries, Fakes, and Reproductions: A Handbook for the Art Dealer and Collector* (New York: Frederick A. Praeger, 1966), xii–xiii.

52. Compare these arguments to those made about the importance of art forgeries: Alfred Lessing, "What Is Wrong with a Forgery?," in *The Forger's Art: Forgery and the Philosophy of Art*, ed. Denis Dutton (Berkeley, Los Angeles, London: University of California Press, 1983).

53. Lessing in Dutton, *The Forger's Art*, 59.

54. Rosen, "The Jilting of Athene," 499.

55. Clark Kerr articulated sentiments shared by many when he noted that, although evidence is indeed "occasionally fabricated, distorted or misrepresented" and "analysis is sometimes faulty," the "system of checks and balances . . . takes good care of these errors"; "unscrupulously ambitious scientists and scholars . . . [are] usually caught and discredited." Kerr, "The Academic Ethic," 143.

56. Mazur, "Allegations of Dishonesty," 178–179.

57. "NIH's Raub on Misconduct," *The Scientist* 1 (15 December 1986): 19. The differences between fields are also discussed in R. E. Kuttner, "Fraud in Science," *Science* 227 (1985): 466.

58. As Evans wrote about accounting fraud in the nineteenth century: "The closeness with which one crime follows upon another, and the similarity of motive that lies at the bottom of them all, will sufficiently show that they do not represent the simple perverseness of individual natures, but are so many indices of a depreciated, and apparently bad, moral atmosphere that has of late pervaded the whole of the commercial world." D. Morier Evans, *Fakes, Failures, and Frauds* (London: Groombridge & Sons, 1859), 5.

59. Patricia Woolf, " 'Pressure to Publish' Is a Lame Excuse for Scientific Fraud," *Chronicle of Higher Education* 34 (23 September 1987): A52.

60. National Academy of Sciences, *On Being a Scientist* (Washington, D.C.: National Academy Press, 1989), 19.

61. Roger H. Davidson, "Subcommittee Government: New Channels for Policy Making," in *The New Congress,* ed. Thomas E. Mann and Norman J. Ornstein (Washington, D.C.: American Enterprise Institute for Public Policy Research, 1981), 117.

62. Legislative oversight includes "review of the actions of federal departments, agencies, and commissions, and of the programs and policies they administer, including review that takes place during program and policy implementation as well as afterward." Joel D. Aberbach, *Keeping a Watchful Eye: The Politics of Congressional Oversight* (Washington, D.C.: The Brookings Institution, 1990), 2. Also see Christopher H. Foreman, Jr., *Signals from the Hill: Congressional Oversight and the Challenge of Social Regulation* (New Haven: Yale University Press, 1988).

63. Barry Bozeman, "A Governed Science and a Self-Governing Science: The Conflicting Value of Autonomy and Accountability," in *Science and Technology Policy: Perspectives and Developments,* ed. Joseph Haberer (Lexington: Lexington Books, 1977), 55–56.

64. Aberbach, *Keeping a Watchful Eye,* 22, 27; Samuel P. Huntington, "Congressional Response to the Twentieth Century," in *The Congress and America's Future,* ed. David B. Truman (Englewood Cliffs: Prentice-Hall, 1965), 20. Under the current rules of the U.S. House of Representatives, this committee has formal responsibility for "reviewing and studying, on

a continuing basis, all laws, programs, and Government activities dealing with or involving non-military research and development." *Rules of the House of Representatives*, Rule X, cl. 3, Sec. 693 (f) and (h).

65. This section is based on research I conducted, with Jeffrey K. Stine, on congressional hearings on science and technology [Marcel C. LaFollette and Jeffrey K. Stine, "Congressional Hearings on Science and Technology Issues: Strengths, Weaknesses, and Suggested Improvements," Report to the Carnegie Commission on Science, Technology, and Government, September 1990]. Our research found a steady increase in such hearings since the 1950s.

66. U.S. Congress, House Committee on Science and Technology, Subcommittee on Investigations and Oversight, *Fraud in Biomedical Research*, Hearings, 31 March 1981, 97th Congress, 1st session (Washington, D.C.: U.S. Government Printing Office, 1981).

67. William J. Broad, "Congress Told Fraud Issue 'Exaggerated'," *Science* 212 (24 April 1981): 421.

68. For example, the U.S. Congress passed the Health Research Extension Act (P.L. 99–158) in 1985, which directed the Department of Health and Human Services to develop regulations and to require grantee institutions to certify that they had established administrative mechanisms to investigate accusations and that they would report "any investigation of suspected fraud which appears substantial."

69. Broad and Wade, *Betrayers of the Truth*. Peter David, "The System Defends Itself," *Nature* 303 (2 June 1983): 369. Also see "Fraud and Secrecy: The Twin Perils of Science," *New Scientist* 98 (9 June 1983): 712–713. The session was held in May 1983 at the annual meeting of the American Association for the Advancement of Science, Detroit, Michigan.

70. One of the other speakers at that session, Daryl E. Chubin, recalls that Norton Zinder was also one of the few scientists willing to appear in such a public debate with Wade. Daryl E. Chubin, conversation with author, June 1991.

71. David, "The System Defends Itself," 369.

72. This section is based on research on the congressional oversight process on science and technology programs, conducted for the Carnegie Commission on Science, Technology, and Government [Marcel C. LaFollette, "Congressional Oversight of Science and Technology Programs," September 1990].

73. *Science and the Congress*, The Third Franklin Conference (Philadelphia: The Franklin Institute Press, 1978), 116.

74. Daniel S. Greenberg, "Fraud Inquiry: NIH on the Capitol Griddle (continued)," *Science & Government Report* 18 (1 May 1988): 4.

75. U.S. Congress, House Committee on Government Operations, Subcommittee on Human Resources and Intergovernmental Relations,

Scientific Fraud and Misconduct and the Federal Response, Hearing, 11 April 1988, 100th Congress, 2d session.

76. U.S. Congress, House Committee on Energy and Commerce, Subcommittee on Oversight and Investigations, *Scientific Fraud and Misconduct in the National Institutes of Health Biomedical Grant Programs,* Hearing, 12 April 1988, 100th Congress, 2d session.

77. Greenberg, "Fraud Inquiry," 6.

78. Mitchell Zuckoff, "Health Agencies Hinder Research-Fraud Probes, House Committee Told," *Boston Globe* 233 (12 April 1988): 12; Pamela S. Zurer, "Research Fraud As Criminal Offense Argued," *Chemical & Engineering News* 66 (18 April 1988): 6.

79. Mark B. Roman, "When Good Scientists Turn Bad," *Discover* (April 1988): 50–58; Paul W. Valentine, "Drug Therapy Researcher Is Indicted," *Washington Post* 111 (16 April 1988): A1, A14; Pamela S. Zurer, "Researcher Criminally Charged with Fraud," *Chemical & Engineering News* 66 (25 April 1988): 5; and David L. Wheeler, "Researcher Is Indicted for Falsifying Data and Impeding Investigation of His Work," *Chronicle of Higher Education* 34 (27 April 1988): A4 and A12.

80. Roman, "When Good Scientists Turn Bad," 53; Eugene Garfield, "The Impact of Scientific Fraud," in *Guarding the Guardians: Research on Peer Review,* Proceedings of the First International Congress on Peer Review in Biomedical Publication, Chicago, 10–12 May 1989 (Chicago: American Medical Association, 1989), reporting on research conducted by the Institute for Scientific Information.

81. Cyril Burt's data on the inheritability of intelligence apparently influenced educational policy in England (Broad and Wade, *Betrayers of the Truth,* 205). Even after his work was discredited, researchers and writers continued to cite it. Diane B. Paul analyzed sections of introductory genetics textbooks published between 1978 and 1984 and found that "nearly half" of them cited or used Burt's data, some even basing their conclusions about genetics or IQ on Burt's work. Diane B. Paul, "The Nine Lives of Discredited Data: Old Textbooks Never Die—They Just Get Paraphrased," *The Sciences* 27 (May 1987): 26–30. Also see Stephen Gould, "The Case of the Creeping Fox Terrier Clone," *Natural History* 97 (January 1988): 16–24.

82. This is not to say that federal involvement in the issue was welcomed warmly by scientists. The editors of the *New England Journal of Medicine,* for example, argued in a June 1988 editorial against establishment of any federal ethics oversight system and stated that they believed that the Congress was simply responding to false impressions of a lack of concern or action by scientists: "The biomedical-research community is willing and able to police itself and is taking steps to do so more effectively. Let us hope that Congress will give this process time to work" (p. 1463). Marcia Angell and Arnold S. Relman, "A Time for Congressional

Restraint," *New England Journal of Medicine* 318 (2 June 1988): 1462–1463.

83. For example, the hearings held 28 June 1989 by the House Committee on Science, Space, and Technology, Subcommittee on Investigations and Oversight. U.S. Congress, *Maintaining the Integrity of Scientific Research*, 101st Congress, 1st Session (Washington, D.C.: U.S. Government Printing Office, 1990).

84. William Booth, "A Clash of Cultures at Meeting on Misconduct," *Science* 243 (3 February 1989): 598.

85. Daniel S. Greenberg, *The Politics of Pure Science* (New York: The New American Library, 1967).

86. David Dickson, *The New Politics of Science* (New York: Pantheon, 1984; Chicago: University of Chicago Press, 1988, with new preface).

87. Michael Levi, *Regulating Fraud: White-Collar Crime and the Criminal Process* (London: Tavistock Publications, 1987), 1–2.

88. Barry Bozeman, "A Governed Science and a Self-Governing Science: The Conflicting Value of Autonomy and Accountability," in Haberer, *Science and Technology Policy*, 55; also see U.S. Congress, House Committee on Science and Technology, Task Force on Science Policy, *The Regulatory Environment for Science*, 99th Congress, 2d session (Washington, D.C.: U.S. Government Printing Office, 1986); also published as Office of Technology Assessment, *The Regulatory Environment for Science* (Washington, D.C.: U.S. Government Printing Office, 1986).

89. Bozeman, "A Governed Science," 56.

CHAPTER 2: CLASSIFYING VIOLATIONS

1. Richard H. Blum, *Deceivers and Deceived: Observations on Confidence Men and Their Victims, Informants and Their Quarry, Political and Industrial Spies and Ordinary Citizens* (Springfield: Charles C. Thomas, 1972), 10.

2. Richard P. Suttmeier, "Corruption in Science: The Chinese Case," *Science, Technology, & Human Values* 10 (Winter 1985): 50.

3. These distinctions are based loosely on classifications found in Charles E. Reagan, *Ethics for Scientific Researchers*, 2d ed. (Springfield: Charles C. Thomas, Publisher, 1971).

4. Suttmeier, "Corruption in Science," points out parallels in the literature on political corruption, where "one of the main methodological problems in studying corruption is to define it" (p. 50).

5. Marcia Angell, remarks in session on "Scientific Fraud and the Pressures to Publish," Council of Biology Editors, Annual Meeting, Raleigh, N.C., 20 May 1986.

6. J. H. Edwards, "Estimation of Burt," *New Scientist* 94 (17 June

1982): 803; cited in Alexander Kohn, *False Prophets: Fraud and Error in Science and Medicine* (New York: Basil Blackwell, 1986), 57.

7. Harry M. Paull, *Literary Ethics: A Study in the Growth of the Literary Conscience* (London: Thornton Butterworth Ltd., 1928), 13.

8. Anthony N. Doob, "Understanding the Nature of Investigations into Alleged Fraud in Alcohol Research: A Reply to Walker and Roach," *British Journal of Addiction* 79 (1984): 169.

9. Ibid., 170.

10. Jonathan L. Fairbanks, "The Art That Needs Another Look," *Yankee Magazine* 53 (November 1989): 84.

11. Ian Haywood, *Faking It: Art and the Politics of Forgery* (London: The Harvester Press, 1987), 8.

12. George Savage, *Forgeries, Fakes, and Reproductions: A Handbook for the Art Dealer and Collector* (New York: Frederick A. Praeger, 1963), 1.

13. Fairbanks, "The Art That Needs," 84.

14. Harold Bennington, *The Detection of Frauds* (Chicago: LaSalle Extension University, 1952), 3.

15. Stephen Lock, *A Difficult Balance: Editorial Peer Review in Medicine* (Philadelphia: ISI Press, 1986), 45.

16. D. Morier Evans, *Fakes, Failures, and Frauds* (London: Groombridge & Sons, 1859), 1.

17. Dennis Rosen, "The Jilting of Athene," *New Scientist* 39 (5 September 1968): 497.

18. Edmund Kerchever Chambers, a scholar at Corpus Christi College, Oxford, made this point eloquently in 1891: "Whatever the motive, there must be at least the intention to deceive, either by attributing to a real or imaginary author work that is not his, or by passing off as record of fact what is mere fiction" (p. 5). Edmund Kerchever Chambers, *The History and Motives of Literary Forgeries, Being the Chancellor's English Essay for 1891* (Folcraft: Folcraft Library Editions, 1975).

19. Alfred Lessing, "What Is Wrong with a Forgery?," in *The Forger's Art: Forgery and the Philosophy of Art*, ed. Denis Dutton (Berkeley, Los Angeles, London: University of California Press, 1983), 64–65.

20. The preface and introduction to Frank Arnau, *The Art of the Faker* (Boston: Little, Brown, 1961), discusses this point.

21. Miriam Milman, *Trompe-l'oeil Painting* (Geneva: Skira, 1982).

22. William A. Thomas, letter to author, March 1989.

23. Ibid.

24. Gerald O'Connor, "The Hoax as Popular Culture," *Journal of Popular Culture* 19 (1976): 767–774.

25. Ibid., 768.

26. Paull, *Literary Ethics*, 13.

27. Some famous hoaxes are described in the following: Stephen Fay,

Lewis Chester, and Magnus Linklater, *Hoax: The Inside Story of the Howard Hughes-Clifford Irving Affair* (New York: The Viking Press, 1972); M. Hirsh Goldberg, *The Book of Lies: Schemes, Scams, Fakes, and Frauds That Have Changed the Course of History and Affected Our Daily Lives* (New York: William Morrow and Company, Inc., 1990); Ian Haywood, *Faking It: Art and the Politics of Forgery* (London: The Harvester Press, 1987); Alexander Klein, *Grand Deception: The World's Most Spectacular and Successful Hoaxes, Impostures, Ruses, and Frauds* (Philadelphia: J. B. Lippincott Company, 1955); William Jay Smith, *The Spectra Hoax* (Middletown, Conn.: Wesleyan University Press, 1962); and Carlson Wade, *Great Hoaxes and Famous Imposters* (New York: Jonathan David Publishers, 1976).

28. Rosen, "The Jilting of Athene," 497.

29. Many years ago, William D. Hahn delighted fellow graduate students at Indiana University with tale after tale of similar pranks played by MIT undergraduates in the 1960s. I "blame" him for my interest in science and engineering hoaxes. Thanks, Bill, wherever you are.

30. Michael Innes, *Picture of Guilt* (New York: Harper & Row, Publishers, 1988), 10. ["Michael Innes" is the pseudonym of novelist James Innes Michael Stewart.]

31. Melvin E. Jahn and Daniel J. Woolf, eds., *The Lying Stones of Dr. Johann Bartholomew Adam Beringer, being his Lithographiae Wirceburgensis* (Berkeley and Los Angeles: University of California Press, 1963): 2.

32. Ibid., 3.

33. Klein, *Grand Deception*.

34. Daniel Webster Hering, *Foibles and Fallacies of Science: An Account of Celebrated Scientific Vagaries* (New York: D. Van Nostrand Company, 1924), 95. Also see Goldberg, *The Book of Lies*.

35. Kim McDonald, "Big Astronomical 'Discovery' Turns Out to Be Interference Signal from Television," *Chronicle of Higher Education* 36 (28 February 1990): A1 and A11; and M. Mitchell Waldrop, "The Puzzling Pulsar That Wasn't There," *Science* 247 (23 February 1990): 910.

36. U.S. Congress, House Committee on Science, Space, and Technology, Subcommittee on Investigations and Oversight, *Maintaining the Integrity of Scientific Research*, 101st Congress, 1st Session (Washington, D.C.: U.S. Government Printing Office, 1990), 236.

37. Kohn, *False Prophets*, ix.

38. John R. Sabine, "The Error Rate in Biological Publications: A Preliminary Survey," *Science, Technology, & Human Values* 10 (1985): 62–69; and John R. Sabine, "Accuracy, Statistics, and Fraud," *Nature* 325 (19 February 1987): 656.

39. Doob, "Understanding the Nature," 173.

40. Ibid., 173.

41. McDonald, "Big Astronomical 'Discovery'," and Waldrop, "The Puzzling Pulsar," 910.

42. This definition has been adapted from that in the *Oxford Dictionary of the English Language*. In the United States, the National Science Foundation and the National Institutes of Health have formulated definitions of research misconduct and these regulations will shape future interpretations of the terms. For now, however, there is wide variation. The American Medical Association's *Manual of Style*, 8th ed. (Baltimore: Williams and Wilkins, 1988), for example, calls fraud and plagiarism "deliberate attempts to deceive" (p. 72). The National Science Foundation had defined "misconduct" as including the "fabrication, falsification, plagiarism, or other serious deviation from accepted practices" in proposing, carrying out, or reporting results from research. Some private institutions have chosen broad definitions; for example, Massachusetts Institute of Technology's "Procedures for Dealing with Academic Fraud in Research and Scholarship" acknowledges that "academic fraud can take many forms, including . . . direct interference with the integrity of the work of others," and the rule admits consideration of such disparate acts as tampering with another's computer files, contaminating another's laboratory specimens, or otherwise altering illicitly the natural course of an experiment (MIT, *Policies and Procedures*, Section 3.51).

43. Daryl E. Chubin, "Research Misconduct: An Issue of Science Policy and Practice," Report to the National Science Foundation (23 July 1984); Daryl E. Chubin, "Misconduct in Research: An Issue of Science Policy and Practice," *Minerva* 23 (Summer 1985), 175–202; and Daryl E. Chubin, "Scientific Malpractice and the Contemporary Politics of Knowledge," in *Theories of Science in Society*, ed. Susan E. Cozzens and Thomas F. Gieryn (Bloomington: Indiana University Press, 1990), 144–163.

44. Rosen, "The Jilting of Athene," 498.

45. C. J. List, "Scientific Fraud: Social Deviance or the Failure of Virtue?," *Science, Technology, & Human Values* 10 (Fall 1985): 27–36.

46. Chambers, *The History and Motives of Literary Forgeries*, 36.

47. Rosen, "The Jilting of Athene," 478. Two new books have raised similar questions about Cyril Burt: R. B. Joynson, *The Burt Affair* (London: Routledge, 1989) and Ronald Fletcher, *Science, Ideology, and the Media: The Cyril Burt Scandal* (New Brunswick, N.J.: Transaction Publications, 1990).

48. Haywood, *Faking It*, 21–29.

49. Wally Herbert, "Did Peary Reach the Pole?," *National Geographic* 174 (September 1988): 387–413.

50. In response to the controversy, the National Geographic Society (which had helped to underwrite the original Peary expedition) supported reexamination of the claim; but the results were inconclusive and the dispute continues. Herbert, "Did Peary Reach the Pole?"; Mark Muro, "A

Geographic Retreat," *Boston Globe* (2 October 1988): A21 and A24; and Eliot Marshall, "Peary's North Pole Claim Reexamined," *Science* 243 (3 March 1989): 1132.

51. Savage, *Forgeries, Fakes*, 1. As Arnau writes, "the gradations between original and fake are many and subtle" [Frank Arnau, *The Art of the Faker: 3,000 Years of Deception* (Boston: Little, Brown and Company, 1961)].

52. David L. Wheeler, "Two Universities Chastised on Fraud Investigations," *Chronicle of Higher Education* 33 (3 June 1987): 1 and 7; and Gregory Byrne, "Breuning Pleads Guilty," *Science* 242 (7 October 1988): 27, reporting on the federal indictment of Stephen Breuning.

53. See Barbara J. Culliton, "The Case of the Altered Notebooks: Part IV," *Science* 248 (18 May 1990): 809, reporting on a 14 May 1990 hearing of the U.S. Congress, House Committee on Energy and Commerce, Subcommittee on Investigations and Oversight. A letter written to Thereza Imanishi-Kari from the NIH's Office of Scientific Integrity (1 February 1990) stated that the NIH investigation focused on "the possibility that substantial portions of the claims related to the immunological aspects of the *Cell* paper were not supported by proper experiments and reliable data at the time of the paper's submission; the possibility that after problems with the paper were brought to light, there was systematic fabrication and falsification of data to support the paper; and the possibility that falsified and fabricated data regarding immunological aspects of the paper were included in representations to the National Institutes of Health and in published letters of correction to the 1986 *Cell* paper." Quoted in Pamela S. Zurer, "Scientific Misconduct: Criminal Inquiry on Scientist Urged," *Chemical & Engineering News* 68 (21 May 1990): 6.

54. Mike Muller, "Why Scientists Don't Cheat," *New Scientist* 74 (2 June 1977): 522–523; William J. Broad and Nicholas Wade, *Betrayers of the Truth: Fraud and Deceit in the Halls of Science* (New York: Simon and Schuster, 1982), 151–153; Kohn, *False Prophets*, 112–114; and Robert J. Gullis, "Statement," *Nature* 265 (1977): 764.

55. For example, Arthur Pierce Middleton and Douglass Adair, "The Mystery of the Horn Papers," in *Evidence for Authorship: Essays on Problems of Attribution*, ed. David V. Erdman and Ephim G. Fogel (Ithaca: Cornell University Press, 1966), 379–386.

56. Arnau, *The Art of the Faker*, describes a number of these, such as Roman coins, medals, jewelry, badges of office, and so forth. Also see Bernard Ashmole, *Forgeries of Ancient Sculpture, Creation and Detection*, The first J. L. Myres Memorial Lecture, delivered in New College, Oxford, on 9 May 1961 (Oxford, England: Holywell Press, 1962); and Anthony Grafton, *Forgers and Critics: Creativity and Duplicity in Western Scholarship* (Princeton, N.J.: Princeton University Press, 1990).

57. Rosen, "The Jilting of Athene," 497.

58. J. S. Weiner, *The Piltdown Forgery* (New York: Oxford University Press, 1955, reprinted by Dover Publications, 1980); Charles Blinderman, *The Piltdown Inquest* (Buffalo: Prometheus Books, 1986); Frank Spencer, *Piltdown: A Scientific Forgery* (New York: Oxford University Press, 1990); and Adolf Reith, *Archaeological Fakes*, translated from the German by Diana Imber (New York: Praeger Publishers, 1970), 34–49.

59. Rosen, "The Jilting of Athene," 498.

60. Weiner, *Piltdown Forgery*; Blinderman, *Piltdown Inquest*; Spencer, *Piltdown*; Nicholas Wade, "Voice from the Dead Names New Suspect for Piltdown Hoax," *Science* 202 (8 December 1978): 1062; Stephen Gould, "The Piltdown Conspiracy," *Natural History* 89 (August 1980): 8–28; and Mary Lukas, "Teilhard and the Piltdown 'Hoax': A playful prank gone too far? Or a deliberate scientific forgery? Or, as it now appears, nothing at all?," *America* 144 (23 May 1981): 424–427.

61. Stephen Gould, "Piltdown Revisited," *Natural History* 88 (March 1979): 88.

62. Rosen, "The Jilting of Athene," 498; Reith, *Archeological Fakes*. Joseph Hixson, *The Patchwork Mouse* (Garden City: Anchor Press/Doubleday, 1976). Another famous deception is described in Arthur Koestler, *The Case of the Midwife Toad* (London: Hutchinson, 1971).

63. Rosen, "The Jilting of Athene," 498.

64. Chapter 7 describes a plausible but fictitious example from *The Affair*, in which a diffraction photo is falsified.

65. Roger Lewin, "Mammoth Fraud Exposed," *Science* 242 (2 December 1988): 1246. See also Jay F. Custer, John C. Kraft, and John F. Wehmiller, "The Holly Oak Shell," *Science* 243 (13 January 1989): 151; and Roger Lewin, "The Holly Oak Shell," *Science* 243 (13 January 1989): 151–152.

66. David Dickson, "Feathers Still Fly in Row over Fossil Bird," *Science* 238 (23 October 1987): 475–476; and Alan J. Charig, Frank Greenaway, Angela C. Milner, Cyril A. Walker, and Peter J. Whybrow, "*Archaeopteryx* Is Not a Forgery," *Science* 232 (2 May 1986): 622–626.

67. Walter Hart Blumenthal, *False Literary Attributions, Works Not Written by Their Supposed Authors, or Doubtfully Ascribed* (Philadelphia: George S. MacManus Company, 1965), p. 21.

68. The novel *Mask of the Jaguar* revolves around a typical deception in archaeology, where dealers carefully plant objects at a site, over several years, so that archaeologists would falsely date a Mayan ruin and thereby increase its value. Jessica North, *Mask of the Jaguar* (New York: Coward, McCann & Geoghegan, 1981).

69. Charles Hamilton, *Great Forgers and Famous Fakes: The Manuscript Forgers of America and How They Duped the Experts* (New York: Crown Publishers, 1980), 257.

70. Similar accusations were made against a well-known journalist in

1989: Eleanor Randolph, "New Yorker Libel Suit Dismissed," *Washington Post* (5 August 1989): C1–C2. Also see "Dangerous Liaisons: Journalists and Their Sources," *Columbia Journalism Review* 28 (July/August 1989): 21–35.

71. Ian St. James-Roberts, "Are Researchers Trustworthy," *New Scientist* (2 September 1976): 481; and Rosen, "The Jilting of Athene," 498.

72. Rosen, "The Jilting of Athene."

73. In 1989, for example, *Nature* reprinted Babbage's original 1830 essay: Charles Babbage, "The Decline of Science in England," *Nature* 340 (17 August 1989): 499–502.

74. William Booth, "Voodoo Science," *Science* 240 (15 April 1988): 274; Wade Davis, "Zombification," *Science* 240 (24 June 1988): 1715; Wade Davis, *Passage of Darkness: The Ethnobiology of a Haitian Zombie* (Chapel Hill: University of North Carolina Press, 1988). In this case, the criticism of Davis appears to have been "relayed" to the scholarly press by third parties, prior to publication. Davis is an anthropologist; his principal critics and accusers in 1988 were a pharmacologist and a toxicologist.

75. Statement made in Chris Raymond, "Study of Patient Histories Suggests Freud Suppressed or Distorted Facts That Contradicted His Theories," *Chronicle of Higher Education* 37 (29 May 1991): A5. Also see Frank Sulloway, "Reassessing Freud's Case Histories: The Social Construction of Psychoanalysis," *Isis* 82 (June 1991): 245–275.

76. Gina Kolata, "Yb or Not Yb? That Is the Question," *Science* 236, (8 May 1987): 663–664.

77. Sabine, "The Error Rate," and Sabine, "Accuracy, Statistics."

78. Stephen Budiansky, "Research Fraud—New Ways of Shading Truth," *Nature* 315 (6 June 1985): 447.

79. Office of Inspector General, *Semiannual Report to the Congress*, No. 4 (1 October 1990–31 March 1991) (Washington, D.C.: National Science Foundation, 1991), 30.

80. Data obtained through Freedom of Information Act request by David L. Wheeler, "U.S. Has Barred Grants to 6 Scientists in Past 2 Years," *Chronicle of Higher Education* 37 (3 July 1991): A1, A6–A7; Office of Scientific Integrity Review, *First Annual Report, Scientific Misconduct Investigations, March 1989–December 1990* (Washington, D.C.: U.S. Department of Health and Human Services, Office of Assistant Secretary for Health, n.d.).

81. Alexander Lindey, *Plagiarism and Originality* (New York: Harper and Brothers, 1952), 2. Peter Shaw, "Plagiary," *American Scholar* 51 (Summer 1982): 327, calls plagiarism "the wrongful taking of and representing as one's own the ideas, words, or inventions of another." Thomas Mallon, *Stolen Words: Forays into the Origins and Ravages of Plagiarism* (New York: Ticknor and Fields, 1989); K. R. St. Onge, *The Melancholy Anatomy of Plagiarism* (Lanham, Md.: The University Press of America,

1988); Morris Freedman, "Plagiarism among Professors or Students Should Not Be Excused or Treated Gingerly," *Chronicle of Higher Education* 34 (10 February 1988): A48; Paull, *Literary Ethics*; and R. Glynn Owens and E. M. Hardley, "Plagiarism in Psychology—What Can and Should Be Done?," *Bulletin of the British Psychological Society* 38 (October 1985): 331–333.

82. Walter Redfern, *Clichés and Coinages* (London: Basil Blackwell, 1989), 67; and Mallon, *Stolen Words*, 6. Johnson's *Dictionary of the English Language* (1755) defined plagiarism as "Theft; literary adoption of the thoughts or works of another" (quoted in Mallon, *Stolen Words*, 11).

83. Redfern, *Clichés*, 67.

84. American Historical Association, "Statement on Plagiarism," *Perspectives* 24 (October 1986): 8; and Paull, *Literary Ethics*, 104.

85. First statement: William Raub, quoted in Alison Bass, "Scientific Community Takes Stock of System that Winks at Plagiarism," *Boston Globe* 234 (2 December 1988): 74. Second statement: scientist quoted in Chris Raymond, "Allegations of Plagiarism of Scientific Manuscript Raise Concerns about 'Intellectual Theft'," *Chronicle of Higher Education* 35 (19 July 1989): A5.

86. Morton Hunt, "A Fraud that Shook the World of Science," *New York Times Magazine* (1 November 1981): 69. Also see Alison Bass, "Plagiarism Topples Leading Scientist," *Boston Globe* 234 (29 November 1988): 1; Alison Bass, "Scientific Community Takes Stock of System that Winks at Plagiarism," *Boston Globe* 234 (2 December 1988): 1 and 74; Alison Bass and Peter Gosselin, "Plagiarism First Reported to Magazine, Not Harvard," *Boston Globe* 234 (30 November 1988): 1 and 26; and Kim McDonald, "Noted Harvard Psychiatrist Resigns Post after Faculty Group Finds He Plagiarized," *Chronicle of Higher Education* 35 (7 December 1988): A1 and A6.

87. George H. Daniels, "Acknowledgment," *Science* (14 January 1972): 124–125. Records of correspondence between historians involved in this case are located in the Smithsonian Institution Archives, Nathan Reingold Collection, General Correspondence, Boxes 2 and 4.

88. Paull, *Literary Ethics*; Lindey, *Plagiarism and Originality*; and Mallon, *Stolen Words*.

89. *A Guide for Wiley-Interscience and Ronald Press Authors in the Preparation and Production of Manuscripts and Illustrations*, 2d ed. (New York: John Wiley & Sons, 1979), 101.

90. Eugene Garfield, "More on the Ethics of Scientific Publication: Abuses of Authorship Attribution and Citation Amnesia Undermine the Reward System of Science," *Current Contents*, No. 30, (26 July 1982): 8. Also see Suttmeier, "Corruption in Science," 53.

91. Garfield, "More on the Ethics."

92. J. A. W. Faidhi and S. K. Robinson, "An Empirical Approach for

Detecting Program Similarity and Plagiarism within a University Programming Environment," *Computers and Education* 11 (1987): 11.

93. John Walsh, "Earthquake Research Center Siting Triggers California Tremors," *Science* 233 (5 September 1986): 1031; Karen Wright, "Earth-shaking Controversy,"*Nature* 323 (9 October 1986): 479; untitled news article on alleged plagiarism in earthquake proposal, *Chronicle of Higher Education* (26 November 1986): 6; and John Walsh, "GAO Finds Fault with NSF Award," *Science* 237 (17 July 1987): 241–242.

94. Michael Gordon, *Running a Referee System* (Leicester, England: Primary Research Publication Centre, University of Leicester, 1983), 7.

95. Lessing in Dutton, *The Forger's Art*, 68.

96. Ibid., 88.

97. Ibid., 75.

98. Ibid.

99. Shaw, "Plaigary"; Mallon, *Stolen Words*; Redfern, *Clichés*; and St. Onge, *Melancholy Anatomy*.

100. International Committee of Medical Journal Editors, "Uniform Requirements for Manuscripts Submitted to Biomedical Journals," *Annals of Internal Medicine* 108 (February 1988): 259.

101. Ibid., 259.

102. Cyril Burt, for example, apparently listed fictitious coauthors. See Hawkes, "Tracing Burt's Descent"; and Broad and Wade, *Betrayers of the Truth*, 203–211.

103. Joel M. Kaufman and Helen E. Plotkin, "Status of Manuscripts," *Chemical & Engineering News* 68 (12 March 1990): 3.

104. Colleen Cordes, "Professor Asserts Stanford Lied to U.S. about Misconduct Probe," *Chronicle of Higher Education* 37 (13 March 1991): A24.

105. On more than one occasion, when serving as a reviewer on grant proposals or on fellowship selection committees, I have read proposals or resumes that listed articles as "accepted" by the journal of which I was editor—when they were not.

106. Gordon, *Running a Referee System*, 17.

107. This case is discussed in Broad and Wade, *Betrayers of the Truth*, 162–180; Kohn, *False Prophets*, 88–91; Hunt, "A Fraud that Shook the World," 42–75; William J. Broad, "Imbroglio at Yale (I): Emergence of a Fraud," *Science* 210 (1 October 1980): 38–41; William J. Broad, "Fraud and the Structure of Science," *Science* 212 (10 April 1981): 137–141; William J. Broad, "Yale Adopts Plan to Handle Charges of Fraud," *Science* 212 (17 April 1981): 311; William J. Broad, "Team Research: Responsibility at the Top," *Science* 213 (3 July 1981): 114–115; and William J. Broad, "Yale Announces Plan to Handle Charges of Fraud," *Science* 218 (1 October 1982): 37.

108. Eliot Marshall, "Should Science Journals Play Cop?," *Science* 248 (13 April 1990): 155.

109. Hawkes, "Tracing Burt's Descent," 674; L. S. Hearnshaw, *Cyril Burt, Psychologist* (London: Hodder and Stoughton, 1979); Broad and Wade, *Betrayers of the Truth*; and Jones, "Obsession," 49.

110. Doob, "Understanding the Nature," 172.

111. Hunt, "A Fraud that Shook the World," 69.

112. John Donne: "Though there were many giants of old in Physics and Philosophy, yet I say with Didacus Stella: a dwarf standing on the shoulder of a giant may see farther than the giant himself." Quoted in Lindey, *Plagiarism and Originality*, 236.

113. Hunt, "A Fraud That Shook the World," 69.

114. Marcia Barinaga, "NIMH Assigns Blame for Tainted Studies," *Science* 245 (25 August 1989): 812.

115. See description of the disposition of the allegations against Philip Berger in Wheeler, "U.S. Has Barred Grants."

116. Leslie Roberts, "The Race for the Cystic Fibrosis Gene," *Science* 240 (8 April 1988): 141; and Leslie Roberts, "Race for the Cystic Fibrosis Gene Nears End," *Science* 240 (15 April 1988): 282–285.

117. Jacqueline Olds and Richard Stanton Schwartz, "The Scientist as Patient," *McLean Hospital Journal* 4, Number 3 (1979): 109.

118. Lindey, *Plagiarism and Originality*, 234. In some cases, the U.S. government has attempted to recover grant money paid for research that was not performed as stated.

119. Lessing, in Dutton, *The Forger's Art*, 65.

120. William A. Thomas, letter to author, May 1987, paraphrasing several law book definitions.

121. William A. Thomas, letter to author, May 1987.

122. Michael Levi, *Regulating Fraud: White-Collar Crime and the Criminal Process* (London: Tavistock Publications, 1987); and Harold Bennington, *The Detection of Frauds*, 3.

123. Michael Heylin, "AAAS Probes Penalty for Scientific Fraud," *Chemical & Engineering News* 68 (5 March 1990): 15–16.

124. Harold Edgar, "Criminal Law Perspectives on Science Fraud," in American Association for the Advancement of Science-American Bar Association, *Project on Scientific Fraud and Misconduct, Report on Workshop Number Three* (Washington, D.C.: American Association for the Advancement of Science, 1989), 139–144.

125. Redfern, *Clichés*, 65.

126. For current law on copyright, see William S. Strong, *The Copyright Book: A Practical Guide*, 3d ed. (Cambridge: The MIT Press, 1990).

127. Lindey, *Plagiarism and Originality*, 232.

128. Mallon, *Stolen Words*, 24.

129. Ibid., 38–39.

130. Haywood, *Faking It*, 21–70, 137.

131. Christopher Small, *The Printed Word: An Instrument of Popularity* (Aberdeen, Scotland: Aberdeen University Press, 1982), 23.

132. David Ladd, "Securing the Future of Copyright: A Humanist Endeavor," Remarks before the International Publishers Association, Mexico City, 13 March 1984; privately printed pamphlet.

133. *Holmes v. Hurst*, 174 U.S. 82, 19 S. Ct. 606, 43 L. Ed. 904 (1898), cited in Lindey, *Plagiarism and Originality*, 281.

134. See Strong, *The Copyright Book*.

135. Lindey, *Plagiarism and Originality*, 294–295.

136. Haywood, *Faking It*, 131.

137. Ladd, "Securing the Future of Copyright," 8; Marcel C. LaFollette, "U.S. Policy on Intellectual Property in R&D: Emerging Political and Moral Issues," in *The Research System in Transition*, ed. Susan E. Cozzens, Peter Healey, Arie Rip, and John Ziman (Dordrecht: Kluwer Academic Publishers, 1990), 125–140.

138. Ladd, "Securing the Future of Copyright," 10.

139. Donald M. Gillmor, Jerome A. Barron, Todd F. Simon, and Herbert A. Terry, *Mass Communication Law: Cases and Comment*, 5th ed. (St. Paul: West Publishing Company, 1990), chap. 1; and Michael Les Benedict, "'Fair Use' of Unpublished Sources," *Perspectives* 28 (April 1990): 1, 9–10, 12–13.

140. Lindey, *Plagiarism and Originality*, 242.

141. *United States Code*, Title 17, section 107. This chapter focuses on the strict interpretation of the law, as it might relate to deceptive practices by an author. When the law was passed, Congress carefully emphasized that the revision was not intended to inhibit quotation for the purpose of scholarship, criticism, or illustration, or in clearly identified summaries of works. *Circular R21, Reproduction of Copyrighted Works by Educators and Librarians*, from the Library of Congress, Copyright Office, describes the law and legislative rationale.

142. AHA, "Statement on Plagiarism," 7.

143. Ibid., 7.

144. *J. D. Salinger v. Random House, Inc., and Ian Hamilton* (U.S. Court of Appeals, 2d Circuit, 1987); and "Paraphrasing Challenged in Copyright Decision," *Scholarly Communication*, No. 8 (Spring 1987): 5 [quoting Carol Rinzler in *Publishers Weekly*, 24 April 1987].

145. "Paraphrasing Challenged in Copyright Decision."

146. In February 1990, the U.S. Supreme Court let stand a 2d Circuit Court of Appeals ruling that could restrict the use of quotes or paraphrased material from unpublished documents, including those deposited in literary archives. [Ruth Marcus and David Streitfeld, "Justices Permit Strict Curbs on Use of Unpublished Writing," *Washington Post* 113 (21 February 1990): A1 and A4.] Although nothing specific in the ruling prohibits scholars from reporting on facts discovered in the course of their research, many legal scholars believe that the ruling could have an inhibiting effect down the line. For discussion of the ruling, see David J. Garrow, "A Law

Defining 'Fair Use' of Unpublished Sources Is Essential to the Future of American Scholarship," *Chronicle of Higher Education* 36 (18 April 1990): A48; Gerald George, "The Fight Over 'Fair Use': When Is It Safe to Quote?" *OAH Newsletter* 18 (August 1990): 10 and 19.

CHAPTER 3: SCIENTIFIC PUBLISHING

1. National Academy of Sciences, *Federal Support of Basic Research in Institutions of Higher Learning* (Washington, D.C.: National Academy of Sciences, 1964), reprinted in *The Politics of American Science, 1939 to the Present*, ed. James L. Penick et al., rev. ed. (Cambridge: The MIT Press, 1972), 5.

2. Harriet Zuckerman and Robert K. Merton, "Patterns of Evaluation in Science: Institutionalisation, Structure, and Functions of the Referee System," *Minerva* 9 (January 1971): 68.

3. David A. Kronick, *A History of Scientific and Technical Periodicals: The Origins and Development of the Scientific and Technical Press, 1665–1790*, 2d ed. (Metuchen: The Scarecrow Press, 1976); and John C. Burnham, "The Evolution of Editorial Peer Review," *Journal of the American Medical Association* 263 (9 March 1990): 1323–1329.

4. Elizabeth L. Eisenstein, *The Printing Press as an Agent of Change: Communications and Cultural Transformations in Early Modern Europe*, vols. 1 and 2 (Cambridge: Cambridge University Press, 1979).

5. Robert Escarpit, *The Book Revolution* (New York: Unesco Publications, 1966).

6. Vannevar Bush, *Science—The Endless Frontier, A Report to the President on a Program for Postwar Scientific Research* (Washington, D.C.: U.S. Government Printing Office, 1945), 3.

7. Irvin Stewart, *Organizing Scientific Research for War* (Boston: Little, Brown and Company, 1948), 288–289; and Harold Relyea, ed., *Striking a Balance: National Security and Scientific Freedom, First Discussions* (Washington, D.C.: American Association for the Advancement of Science, 1985), 9.

8. J. Merton England, *A Patron for Pure Science: The National Science Foundation's Formative Years, 1945–57* (Washington, D.C.: National Science Foundation, 1982), 175.

9. Discussed in "Role of Professional Societies, Commercial Organizations, Universities, and Government Agencies in the Information Transfer Process" (NSB-71-263) and in U.S. Congress, House Committee on Science and Technology, *The National Science Board: Science Policy and Management for the National Science Foundation, 1968–1980*, 98th Congress, 1st Session (Washington, D.C.: U.S. Government Printing Office, 1983), 522.

10. England, *Patron*, 176–177. By 1954, the Scientific Information

Office had become the Office of Scientific Information. It did not assume a strong role until Congress passed the National Defense Education Act of 1958, which included establishment of a Science Information Service and a Science Information Council to coordinate activities at the National Science Foundation, Library of Congress, and elsewhere in the federal government. Dorothy Schaffter, *The National Science Foundation* (New York: Frederick Praeger, 1969), 126–127.

The NSF Office of Science Information Services was created in 1958. It is said to be the result of recommendations of a panel of the President's Science Advisory Council, chaired by William O. Baker, on whether the government should set up a central scientific information system. In 1959, the Federal Council on Science and Technology directed the NSF to coordinate federal activities, and the NSF responded by increasing the responsibilities of the Science Information Service. Jerome B. Wiesner, *Where Science and Politics Meet* (New York: McGraw-Hill, 1965), 157–158; Harvey A. Averch, *Strategic Analysis of Science and Technology Policy* (Baltimore: Johns Hopkins University Press, 1985), 102–104.

11. Shaffter, *National Science Foundation*, 131.

12. Wiesner, *Where Science and Politics Meet*, 154–155.

13. For example, with COSATI and the NAS-NAE Committee on Scientific and Technical Communication.

14. Herbert C. Morton and Anne Jamieson Price, "Views on Publications, Computers, Libraries," *Scholarly Communication*, No. 5 (Summer 1986): 1–15; and Herbert C. Morton and Anne J. Price, with Robert Cameron Mitchell, *The ACLS Survey of Scholars—Final Report* (Washington, D.C.: American Council of Learned Societies, 1989).

15. William J. Broad, "Crisis in Publishing: Credit or Credibility," *BioScience* 32 (September 1982): 646.

16. Jack G. Goellner, "The Other Side of the Fence: Scholarly Publishing as Gatekeeper," *Book Research Quarterly* 4 (Spring 1988): 18.

17. Robert V. Schnucker, "Journal Editors Speak Out—1. 'I'll Read It in the Library': A Lament and a Sales Pitch," *Scholarly Communication*, No. 3 (Winter 1986): 8.

18. George Lundberg, speaking to CBE Annual Meeting 1989, quoted in "Annual Meeting Reports–1989," *CBE Views* 12 (October 1989): 81.

19. Morton and Price, *The ACLS Survey*.

20. Goellner, "The Other Side," 18.

21. Ibid., 18.

22. Michael Les Benedict, "Historians and the Continuing Controversy over Fair Use of Unpublished Manuscript Materials," *American Historical Review* 91 (October 1986): 859.

23. Eugene Garfield, "More on the Ethics of Scientific Publication: Abuses of Authorship Attribution and Citation Amnesia Undermine the Reward System of Science," *Current Contents*, No. 30 (26 July 1982): 9.

24. Denis Rosen, "The Jilting of Athene," *New Scientist* 39 (5 September 1968): 500.

25. Frederick Bowes III, "Keynote Address," *Proceedings of the Society for Scholarly Publishing* 11 (Washington: Society for Scholarly Publishing, 1990), 1.

26. David Van Tassel, "From Learned Society to Professional Organization: The American Historical Association, 1884–1900," *American Historical Review* 89 (October 1984): 930.

27. "An editor shall not be subject to, nor obligated to accept, any recommendation or instruction from the ACS Board of Directors, Executive Director, Director of Books and Journals, or any other officer or employee of the Society regarding the selection of articles for publication or the contents thereof." *Charter, Constitution, Bylaws, and Regulations of the ACS*, as revised through 1 January 1987, Bulletin 5 (Section R, IX, 2, C), p. 33.

28. American Chemical Society, *Handbook for Members* (1987), "Guidelines for Journal Monitoring by the Society Committee on Publications." American Chemical Society, *Charter, Constitutions, Bylaws, and Regulations of the ACS*, as revised through 1 January 1987, Bulletin 5 (Section R IX, 2, "Rights and Responsibilities of Editors of ACS Publications"). Barbara G. Wood, "Monitoring the Health of ACS Publications," *Chemical & Engineering News* 65 (12 October 1987), 24.

29. Goellner, "The Other Side," 15.

30. Edward J. Huth, speech to the annual meeting of the Council of Biology Editors, as reported in *CBE Views* 12 (October 1989): 81.

31. Simon Mitton, "Publishing Industry Mergers: They May Be Bad for Science," *The Scientist* 2 (30 May 1988): 16.

32. Ibid.

33. Alan Singleton, *Societies and Publishers: Hints on Collaboration in Journal Publishing* (Leicester: Primary Communications Research Centre, University of Leicester, 1980).

34. Bowes, "Keynote Address," 2.

35. Marjorie Sun, "Barred Researcher Publishes a Paper," *Science* 219 (21 January 1983): 270.

36. Jonathan Rhoads, speaking in a session on "Who Determines Journal Policy?" at the Council of Biology Editors annual meeting, as reported in "Annual Meeting Reports–1989," *CBE Views* 12 (October 1989): 81.

37. Bowes, "Keynote Address," 1–2.

38. Ken Kalfus, "Scientists Balk at Soaring Journal Prices," *The Scientist* 3 (24 July 1989): 16.

39. Philip H. Abelson, "Support of Scientific Journals," *Science* 214 (23 October 1981).

40. Data given in Mary E. Curtis, "On the Care and Nurturing of Authors," *SSP Letter* 10 (1988): 1–3.

41. John Walsh, "Nonprofits Could Face a Taxing Time," *Science* 241 (19 August 1988): 896; also see *1989 Annual Report*, American Association for the Advancement of Science.

42. Nonprofit tax status is justified by professional organizations' avowed missions of disseminating knowledge related to some educational, scientific, or cultural goal, and it represents a form of federal subsidy for these activities. Other indirect subsidies include preferential postage rates for the publications of qualifying nonprofit educational, cultural, and scientific organizations.

43. Dana L. Roth, "Journal Price Increases," *Science* 238 (30 October 1987): 597.

44. Keith Bowker, "Journal Price Increases," *Science* 238 (30 October 1987): 597.

45. F. F. Clasquin and Jackson B. Cohen, "Prices of Physics and Chemistry Journals," *Science* 197 (29 July 1977): 432.

46. Robert K. Peet, "Collections in the Future Will Be Electronic," *The Scientist* 4 (30 April 1990): 17 and 19.

47. Data from Kalfus, "Scientists Balk"; and Constance Holden, "Libraries Stunned by Journal Price Increases," *Science* 236 (22 May 1987): 908–909.

48. Data from Kalfus, "Scientists Balk," 1 and 16.

49. Holden, "Libraries Stunned." Some critics cited the differential price rise between those journals published in the U.S. (up 10% in 1987 from previous year) and the worldwide average price increase (up 14 to 18% in 1987).

50. Broad, "Crisis in Publishing," 647.

51. Ibid.

52. Abigal Grissom, "Flow of Scientific Progress Creates Wave of New Journals," *The Scientist* 4 (12 November 1990): 17.

53. Katina Strauch, "The Irresistible Force Meets the Immovable Object," *Proceedings of the Society for Scholarly Publishing* 11 (Washington, D.C.: Society for Scholarly Publishing, 1990).

54. Hendrik Edelman, "Quality Management and Public Accountability," *Proceedings of the Society for Scholarly Publishing* 11 (Washington, D.C.: Society for Scholarly Publishing, 1990).

55. For example, Eric Ashby, "The Academic Profession," *Minerva* 7 (January 1970): 90–99; André Cournand, "On the Codes of Science and Scientists," in *From Theoretical Physics to Biology*, ed. M. Marois (Basel: S. Karger, 1973), 436–450; André Cournand and Michael Meyer, "The Scientist's Code," *Minerva* 14 (Spring 1976): 79–96; André Cournand, "The Code of the Scientist and Its Relationship to Ethics," *Science* 198 (18 November 1977): 699–705; and Clark Kerr, "The Academic Ethic and University Teachers: A 'Disintegrating Profession'?," *Minerva* 27 (Summer–Autumn 1989): 139–156.

56. Stephen Fay, Lewis Chester, and Magnus Linklater, *Hoax: The Inside Story of the Howard Hughes-Clifford Irving Affair* (New York: The Viking Press, 1972), 94.

57. Ibid., 94.

58. Chris Raymond, "Allegations of Plagiarism of Scientific Manuscript Raise Concerns about 'Intellectual Theft'," *Chronicle of Higher Education* 35 (19 July 1989): A7.

59. William H. Glaze, "Peer Review: A Foundation of Science," *Environmental Science & Technology* 22 (March 1988), 235.

60. Kerr, "The Academic Ethic," 142.

61. Clark Kerr calls this "the obligation to be skeptical of what is not fully proven." Kerr, "The Academic Ethic," 142.

62. Duncan Lindsey, *The Scientific Publication System in Social Science* (San Francisco: Jossey-Bass Publishers, 1978), 103.

63. Nathan Reingold, interview with the author, 5 August 1987.

64. Related to Merton's norm of "disinterestedness," Cournand's "unselfish engagement," or Hagstrom's "norm of service." Cournand, "The Code of the Scientist," 699–705; Warren O. Hagstrom, *The Scientific Community* (Carbondale: Southern Illinois University Press, 1965).

65. William P. Sisler, "Loyalties and Royalties," *Book Research Quarterly* 4 (Spring 1988): 13.

66. Diana Crane, *Invisible Colleges* (Chicago: University of Chicago Press, 1972), 4; the latter quote is a comment from William A. Thomas, letter to author, May 1987.

67. Michael Gordon, *Running a Referee System* (Leicester: Primary Communications Research Centre, University of Leicester, 1983), 7.

CHAPTER 4: AUTHORSHIP

1. Ian Haywood, *Faking It: Art and the Politics of Forgery* (London: The Harvester Press, 1987), 4.

2. W. S. Gilbert and Sir Arthur Sullivan, lyrics to "Things are Seldom What They Seem," *H.M.S. Pinafore*.

3. Erwin O. Smigel and H. Laurence Ross, "Factors in the Editorial Decision," *American Sociologist* 5 (February 1970): 19–21.

4. Donald Kennedy, "On Academic Authorship," Statement to members of the Stanford University Academic Council (Palo Alto: Stanford University, 1985).

5. "Must Plagiarism Thrive?," *British Medical Journal* 281 (5 July 1980): 41–42.

6. Ibid., 42.

7. Stephen Lock, *A Difficult Balance: Editorial Peer Review in Medicine* (Philadelphia: ISI Press, 1986), 23.

8. "Must Plagiarism Thrive?," 42.

9. Kennedy, "On Academic Authorship," 2.

10. Quoted in Barbara J. Culliton, "Authorship, Data Ownership Examined," *Science* 242 (4 November 1988): 658.

11. Carl Djerassi, *Cantor's Dilemma* (New York: Doubleday, 1989), 51.

12. Many scientists publicly express concern about this problem [for example, R. L. Strub and F. W. Black, "Multiple Authorship," *Lancet* (13 November 1976): 1090–1091].

13. William J. Broad, "The Publishing Game: Getting More for Less," *Science* 211 (13 March 1981): 1137–1139.

14. Data from Strub and Black, "Multiple Authorship," and from Broad, "The Publishing Game," 1137–1139.

15. Broad, "The Publishing Game," 1137.

16. Derek Price, "Multiple Authorship," *Science* 212 (1981): 986.

17. Bernard Dixon, "Journal Editors and Co-authors," *The Scientist* 1 (9 February 1987): 11–12.

18. Francisco J. Ayala, "For Young Scientists, Questions of Protocol and Propriety Can Be Bewildering," *Chronicle of Higher Education* 36 (22 November 1989): 36.

19. Harriet Zuckerman, "Patterns of Name Ordering among Authors of Scientific Papers: A Study of Social Symbolism and Its Ambiguity," *American Journal of Sociology* 74 (1968): 276–291.

20. Daniel S. Greenberg, "Scientific Fraud," *American Scientist* 72 (January–February 1984): 118.

21. William J. Broad, "Crisis in Publishing: Credit or Credibility," *BioScience* 32 (September 1982): 646.

22. Eliot Marshall, "Textbook Credits Bruise Psychiatrists' Egos," *Science* 235 (20 February 1987): 835–836.

23. Gerald L. Klerman, "Textbook Dispute," *Science* 236 (8 May 1987): 655.

24. Marshall, "Textbook Credits," 835.

25. David L. Wheeler, "Researchers Need to Shoulder Responsibility for Detecting Misconduct, Scientist Warns," *Chronicle of Higher Education* 33 (25 February 1987): 13. These are Wheeler's words not Thier's.

26. Broad, "Crisis in Publishing," 645.

27. Broad, "The Publishing Game," 1137.

28. Ravi K. Agrawal, "Cheating in Research," *Chemical & Engineering News* 65 (10 August 1987): 4–5.

29. Mike Muller, "Why Scientists Don't Cheat," *New Scientist* 74 (2 June 1977): 522–523; William J. Broad and Nicholas Wade, *Betrayers of the Truth: Fraud and Deceit in the Halls of Science* (New York: Simon and Schuster, 1982), 151–153; and Alexander Kohn, *False Prophets: Fraud and Error in Science and Medicine* (New York: Basil Blackwell, 1986), 112–114.

30. George Holt Colt, "Too Good to be True," *Harvard Magazine* (July–August 1983): 28.

31. Robert L. Engler, James W. Covell, Paul J. Friedman, Philip S. Kitcher, and Richard M. Peters, "Misrepresentation and Responsibility in Medical Research," *New England Journal of Medicine* 317 (26 November 1987): 1383–1389; "The Morality of Scientists: II—Report of the Faculty *ad hoc* Committee to Investigate Research Fraud," *Minerva* 25 (Winter 1987): 502–512.

32. Eliot Marshall, "San Diego's Tough Stand on Research Fraud," *Science* 234 (31 October 1986): 534.

33. Ibid., 534.

34. "The Morality of Scientists: II," 508.

35. Ibid., 509–511.

36. Ibid., 509.

37. Ibid., 509.

38. Ibid., 512; the forgery is also mentioned in Rex Dalton, "Fraudulent Papers Stain Co-Authors," *The Scientist* 1 (18 May 1987): 8.

39. Dalton, "Fraudulent Papers," 8.

40. Roger P. Croll, "The Non-Contributing Author: An Issue of Credit and Responsibility," *Perspectives in Biology and Medicine* 27 (Spring 1984): 401–407.

41. Ibid., 403.

42. Haywood, *Faking It*, 106.

43. Anthony Bailey, "A Young Man on Horseback," *The New Yorker* 66 (5 March 1990): 56.

44. Ibid., 57.

45. Ibid., 58.

46. Ibid., 59.

47. Haywood, *Faking It*, 106–108.

48. "Ethical Principles for Psychologists," *American Psychologist* 36 (June 1981): 637.

49. International Committee of Medical Journal Editors, "Uniform Requirements for Manuscripts Submitted to Biomedical Journals," *Annals of Internal Medicine* 108 (February 1988): 259.

50. Ellen K. Coughlin, "Concerns about Fraud, Editorial Bias Prompt Scrutiny of Journal Practices," *Chronicle of Higher Education* 35 (15 February 1989): A6.

51. Described in Stephen Budiansky, "Research Fraud—New Ways of Shading Truth," *Nature* 315 (6 June 1985): 447.

52. Barbara Mishkin, testimony before the House Committee on Science, Space, and Technology, Subcommittee on Investigations and Oversight, 28 June 1989. Reprinted in U.S. Congress, *Maintaining the Integrity of Scientific Research*, 101st Congress, 1st session (Washington, D.C.: U.S. Government Printing Office, 1990), 203.

53. Eugene Garfield, "More on the Ethics of Scientific Publication: Abuses of Authorship Attribution and Citation Amnesia Undermine the Reward System of Science," *Current Contents*, No. 30 (26 July 1982): 7, n. 18.

54. Dixon, "Journal Editors."

55. Croll, "The Non-Contributing Author," 404.

56. Ibid., 404.

57. Ibid., 401 and 404.

58. "Information for Contributors," *Science* 251 (4 January 1991): 35–37.

59. Roger Lewin, "The Case of the 'Misplaced' Fossils," *Science* 244 (21 April 1989): 277–279.

60. John A. Talent, "The Case of the Peripatetic Fossils," *Nature* 338 (20 April 1989): 613–615; A. D. Ahluwalia, "Upper Palaeozoic of Lahul-Spiti," *Nature* 341 (7 September 1989): 13–14; S. B. Bhatia, "Early Devonian Ostracodes," *Nature* 341 (7 September 1989): 15; Udai K. Bassi, "The Kinnaur Region," *Nature* 341 (7 September 1989): 15–16; and Phillippe Janvier, "Breakdown of Trust," *Nature* 341 (7 September 1989): 16.

61. Lewin, "The Case of the 'Misplaced' Fossils," 279.

62. Ibid., 279.

63. Vishwa Jit Gupta, "The Peripatetic Fossils: Part 2," *Nature* 341 (7 September 1989): 11–12; Ahluwalia, "Upper Palaeozoic," 13–14; Bhatia, "Early Devonian," 15; Bassi, "The Kinnaur Region," 15–16; Janvier, "Breakdown of Trust," 16.

64. Heinrich K. Erben, "Carelessness, or Good Faith?" *Science* 245 (15 September 1989): 1165–1166.

65. John A. Talent, "The 'Misplaced' Fossils," *Science* 246 (10 November 1989): 740.

66. Ibid., 740.

67. Marcia Barinaga, "NIMH Assigns Blame for Tainted Studies," *Science* 245 (25 August 1989): 812.

68. Derek de Solla Price, "Multiple Authorship," *Science* 212 (1981), 986.

69. Patricia Woolf, remarks to the American Association for the Advancement of Science, Annual Meeting, Chicago, February 1987.

70. David L. Wheeler, "One-Third of Scientists Surveyed at Major University Suspected a Colleague of Fraudulent Research," *Chronicle of Higher Education* 34 (9 September 1987): A6–A7.

71. Joel Tellinghuisen, "The Most Honest Game in Town," *Chemical & Engineering News* 65 (1 June 1987): 2–3.

72. Ibid., 2–3.

73. D. E. Whitney, "Scientific Fraud," *American Scientist* 72 (January–February 1987): 116.

74. Garfield, "More on the Ethics," 6.

75. Croll, "Non-Contributing Authors," 405.

76. Richard H. Blum, *Deceivers and Deceived; Observations on Confidence Men and Their Victims, Informants and Their Quarry, Political and Industrial Spies and Ordinary Citizens* (Springfield: Charles C. Thomas Publisher, 1972), 3.

77. Ibid., 9.

78. Ibid., 8.

79. Thomas Mallon, *Stolen Words: Forays into the Origins and Ravages of Plagiarism* (New York: Ticknor and Fields, 1989), 164.

80. Blum, *Deceivers and Deceived*, 9–10.

81. Ibid., 10.

82. Ibid., 10.

83. Ibid., 10.

84. Ibid., 10.

85. George B. Lundberg and Annette Flanagin, "New Requirements for Authors," *Journal of the American Medical Association* 262 (13 October 1989): 2004.

86. Croll, "Non-Contributing Authors," 404–405; Garfield, "More on the Ethics," nn. 12 and 13; Edward Diener and Rick Crandall, *Ethics in Social and Behavioral Research* (Chicago: University of Chicago Press, 1978); Kennedy, "On Academic Authorship," 3.

CHAPTER 5: DECISION MAKING

1. Quoted in George Savage, *Forgeries, Fakes and Reproductions: A Handbook for the Art Dealer and Collector* (New York: Frederick A. Praeger, Publishers, 1963).

2. Jonathan Yardley, "Closing the Book on Old-Style Publishing," *Washington Post* (27 November 1989): B2.

3. A. Scott Berg, *Max Perkins* (London: Macdonald Futura Publishers, 1981), 3–4.

4. Ibid., 4.

5. Ibid., 4.

6. Quoted in Paul F. Parsons, "The Editorial Committee: Controller of the Imprint," *Scholarly Publishing* 20 (July 1989): 240.

7. Diana Crane, "The Gatekeepers of Science: Some Factors Affecting the Selection of Articles for Scientific Journals," *American Sociologist* 2 (November 1967): 195–201. Also see Jack G. Goellner, "The Other Side of the Fence: Scholarly Publishing as Gatekeeper," *Book Research Quarterly* 4 (Spring 1988): 18.

8. Goellner, "The Other Side," 19.

9. Morey D. Rothberg, "'To Set a Standard of Workmanship and Compel Men to Conform to It': John Franklin Jameson as Editor of the

American Historical Review," *American Historical Review* 89 (October 1984): 957–975.

10. Rothberg, "To Set a Standard," 970.

11. Quoted in Rick Crandall, "Peer Review: Improving Editorial Performance," *BioScience* 36 (October 1986): 608.

12. Jacques Barzun, *On Writing, Editing, and Publishing: Essays Explicative and Hortatory,* 2d ed. (Chicago: University of Chicago Press, 1986), chap. 8 ("Behind the Blue Pencil: Censorship or Creeping Creativity?") and chap. 9 ("Paradoxes of Publishing").

13. Ibid., 103–112.

14. Ibid., 117.

15. Crandall, "Peer Review," 607.

16. Stephen Lock, *A Difficult Balance: Editorial Peer Review in Medicine* (Philadelphia: ISI Press, 1986), xv and 6.

17. Jesse O. Cavenar, "Textbook Dispute," *Science* 236 (8 May 1987): 656, describes one view of the responsibilities of an editor.

18. Parsons, "The Editorial Committee."

19. William H. Glaze, "New Directions: 1990," *Environmental Science & Technology* 23 (November 1989): 1311; and William H. Glaze, "Changes at ES&T," *Environmental Science & Technology* 23 (January 1990): 3.

20. Richard A. Knox, "Doctor-Editor-Patient," *New England—The Boston Sunday Globe* (16 June 1977): 31.

21. Philip Abelson, "Mechanisms for Evaluating Scientific Information and the Role of Peer Review," *Journal of the American Society for Information Science* 41 (1990): 218.

22. George Lundberg, "Appreciation to Our Peer Reviewers," *Journal of the American Medical Association* 251 (10 February 1984): 758.

23. Crandall, "Peer Review," 607.

24. Described in Lawrence K. Altman, "Peer Review Is Challenged," *New York Times* (25 February 1986): C3.

25. Knox, "Doctor-Editor-Patient," 31.

26. Lewin: "The professional skepticism of reviewers is our sole defense." Benjamin Lewin, "Fraud in Science: The Burden of Proof," *Cell* 48 (16 January 1987): 1.

27. Monica M. Creamer, "Build a Cathedral," *Environmental Science & Technology* 21 (April 1987): 315.

28. Ibid., 315.

29. Ken Kalfus, "Scientists Balk at Soaring Journal Prices," *The Scientist* 3 (24 July 1989): 16.

30. Gerald W. Bracey, "The Time Has Come to Abolish Research Journals: Too Many Are Writing Too Much about Too Little," *The Chronicle of Higher Education* (25 March 1987): 44.

31. Quoted in Knox, "Doctor-Editor-Patient," 31.

32. "The Journal's Peer-Review Process," *New England Journal of Medicine* 321 (21 September 1989): 838.

33. Lundberg, "Appreciation," 758.

34. Russell F. Christman, "Manuscript Processing Time," *Environmental Science & Technology* 17 (1983): 161a.

35. William H. Glaze, "Needed: A Speedier Time to Publication," *Environmental Science & Technology* 22 (February 1988): 115.

36. Ibid., 115.

37. Kim McDonald, "Race to Produce Superconducting Materials Stirs Concern over Publications and Patents," *Chronicle of Higher Education* (29 April 1987): 7.

38. Astronomy, 90.2 percent; chemistry, 72 percent; biochemistry and genetics, 65 percent; physics, 61.5 percent; anthropology, 46.5 percent; psychology, 18.8 percent; and sociology, 14 percent. G. Christopher Anderson, "Getting Science Papers Published: Where It's Easy, Where It's Not," *The Scientist* 2 (19 September 1988): 26–27. Anderson cites data from Lowell Hargens in *American Sociological Review* 53(1988): 139–151, and Helmut Abt in *Publications of the Astronomical Society of the Pacific* 100 (1988): 506–508.

39. Constance Holden, "Libraries Stunned by Journal Price Increases," *Science* 236 (22 May 1987): 909.

40. Marcel C. LaFollette, "Journal Peer Review and Public Policy," *Science, Technology, & Human Values* 10 (Winter 1985): 3–5.

41. Lundberg, "Appreciation." *Nature* employs about five technical consultants, who review articles on topics for which there is perceived to be "insufficient expertise" in the editorial office. Bernard Dixon, "John Maddox Offers Surprising Insights into His *Nature*," *The Scientist* 2 (13 June 1988): 13.

42. David Thelen, "The Profession and the Journal of American History," *Journal of American History* 73 (June 1986): 14.

43. Claude Liébecq, "Privileged Communication," *Science* 211 (23 January 1981): 336.

44. Lundberg, "Appreciation," 758.

45. Unsigned essay, "It's Over, Debbie," *Journal of the American Medical Association* 259 (8 January 1988). The case is described in Mark Bloom, "Article Embroils JAMA in Ethical Controversy," *Science* 239 (11 March 1988): 1235–1236.

46. Bloom, "Article Embroils JAMA," 1235.

47. See, for example, Sissela Bok, *Secrets: On the Ethics of Concealment and Revelation* (New York: Pantheon Books, 1982); Bruce Landesman in David Luban, ed., *The Good Lawyer: Lawyers' Roles and Lawyers' Ethics* (Totowa, N.J.: Rowman and Allenheld, 1984); and "To Tell or Not to Tell: Conflicts about Confidentiality," *Report from the Center for Philosophy and Public Policy* 4 (Spring 1984): 1–5.

48. Franz J. Ingelfinger, "Peer Review in Biomedical Publication," *American Journal of Medicine* 56 (May 1974): 691.

49. Ibid., 686.

50. As Lock points out, the author is simply seeking publication in this journal; rejection by one journal does not preclude eventual publication in another. Lock, *A Difficult Balance*, 106.

51. Ingelfinger, "Peer Review," 686.

52. Igor B. Dawid, Maxine F. Singer, and Frances R. Zwanzig, "*PNAS* Reviewing Procedures," *Science* 243 (17 March 1989): 1419.

53. Ingelfinger, "Peer Review."

54. Dixon, "John Maddox Offers," 13, reports that *Nature* often overrides referees' recommendations because, as John Maddox notes, "nearly twice as many papers are approved by both referees as we can possibly accommodate."

55. Ingelfinger, "Peer Review," 687.

56. John Ziman, *Public Knowledge: The Social Dimension of Knowledge* (Cambridge: Cambridge University Press, 1968), 111.

57. For example, Dixon, "John Maddox Offers," 13, cites *Nature's* practice of not compensating referees.

58. Lock, *A Difficult Balance*, 2.

59. Ibid., 21.

60. Ibid., 21.

61. Drummond Rennie, "Guarding the Guardians: A Conference on Editorial Peer Review," *Journal of the American Medical Association* 256 (7 November 1986): 2391.

62. Ibid., 2391.

63. Rennie, "Guarding the Guardians," and LaFollette, "Journal Peer Review."

64. Nathan Reingold, "On Not Doing the Papers of Great Scientists," *British Journal of the History of Science* 20 (1987): 33.

65. Harriet Zuckerman and Robert K. Merton, "Patterns of Evaluation in Science: Institutionalisation, Structure, and Functions of the Referee System," *Minerva* 9 (January 1971): 100.

66. Joel Tellinghuisen, "The Most Honest Game in Town," *Chemical & Engineering News* 65 (1 June 1987): 3; and Bracey, "The Time Has Come," 44.

67. Data on *AJPH* from Alfred Yankauer, "Who Are the Journal Reviewers and How Much Do They Review?," in *Guarding the Guardians: Research on Peer Review*, Proceedings of the First International Congress on Peer Review in Biomedical Publication, Chicago, 10–12 May 1989. Data on *British Medical Journal* from Jane M. Smith and Stephen P. Lock, "The Community of Referees," in *Guarding the Guardians;* and also Lock, *A Difficult Balance*.

68. Denis Dutton, *The Forger's Art: Forgery and the Philosophy of*

Art (Berkeley, Los Angeles, London: University of California Press, 1983), 34–35.

69. Ibid., 35.

70. Ibid., 19.

71. Michael Gordon, *Running a Referee System* (Leicester, England: PRPC, 1983), 7.

72. Lundberg, "Appreciation," 758.

73. Bloom, "Article Embroils JAMA," 1236.

74. See discussion in Ellen K. Coughlin, "Concerns about Fraud, Editorial Bias Prompt Scrutiny of Journal Practices," *Chronicle of Higher Education* 35 (15 February 1989): A4–A7.

75. Robert G. Martin, "Quality of Biomedical Literature," *Science* 235 (9 January 1987): 144.

76. G. H. Freeman, "Accuracy, Statistics and Fraud," *Nature* 325 (19 February 1987): 656.

77. Patricia Woolf, "'Pressure to Publish' Is a Lame Excuse for Scientific Fraud," *Chronicle of Higher Education* 34 (23 September 1987): A52.

78. Alison Bass, "Scientific Community Takes Stock of System that Winks at Plagiarism," *Boston Globe* 234 (2 December 1988): 74.

79. Dale B. Baker, "Recent Trends in Chemical Literature Growth," *Chemical & Engineering News* 59 (1 June 1981): 29–34.

80. Lewin, "Fraud in Science," 1.

81. John Maddox, quoted in Dixon, "John Maddox Offers," 13–14.

82. Stephen Gould, "Piltdown Revisited," *Natural History* 88 (March 1979): 92–95.

83. Brian Evans and Bernard Waite, *IQ and Mental Testing: An Unnatural Science and Its Social History* (Atlantic Highlands: Humanities Press, 1981), 162.

84. Ibid., 162 and 219 (fn. 49).

85. Nigel Hawkes, "Tracing Burt's Descent to Scientific Fraud," *Science* 205 (17 August 1979): 673–675; F. L. Jones, "Obsession Plus Pseudo-Science Equals Fraud: Sir Cyril Burt, Intelligence, and Social Mobility," *Australian and New Zealand Journal of Sociology* 16 (March 1980): 48–55; Broad and Wade, *Betrayers of the Truth,* 79–82, 203–211; and Kohn, *False Prophets,* 52–57.

86. Hawkes, "Tracing Burt's Descent," 674.

87. Kohn, *False Prophets,* 53, describing Kamin's analysis of Burt.

88. Jones, "Obsession Plus Pseudo-Science," 51–52.

89. Bass, "Scientific Community Takes Stock," 74. Also see Alfred Lessing in Dutton, *The Forger's Art,* 63.

90. George Howe Colt, "Too Good to be True," *Harvard Magazine* (July–August 1983): 27.

91. Stephen Budiansky, "Research Fraud—New Ways of Shading Truth," *Nature* 315 (6 June 1985): 447.

92. Lewin, "Fraud in Science," 1–2, describes fraud detected by a referee.

93. Lock, *A Difficult Balance*, 48.

94. Ibid., 48, citing Garth J. Thomas in *Behavioral and Brain Sciences* 5 (June 1982): 240.

95. See Relman's response to Mark Jameson, "Editorial Review," *New England Journal of Medicine* 307 (1982): 849.

96. Gordon, *Running a Referee System*, 17.

97. Lock, *A Difficult Balance*, 34; and John Maddox, "Privacy and the Peer-Review System," *Nature* 312 (1984): 497.

98. Tellinghuisen, "The Most Honest Game," 2. Other people have argued the opposite. Gordon, for example, describes the case against anonymous peer review, but he does worry about referee abuse (Gordon, *Running a Referee System*, 17).

99. Crane, "The Gatekeepers of Science," 196.

100. Tellinghuisen, "The Most Honest Game."

101. Lock, *A Difficult Balance*, 122.

102. Daryl E. Chubin and Edward J. Hackett, *Peerless Science: Peer Review and U.S. Science Policy* (Albany: State University of New York Press, 1990) fills a significant gap in the literature.

103. Marcel C. LaFollette, "On Fairness and Peer Review," *Science, Technology, & Human Values* 8 (Fall 1983): 3–5.

104. Allan Mazur, "Allegations of Dishonesty in Research and Their Treatment by American Universities," *Minerva* 27 (Summer–Autumn 1989): 184–185. Also see William Broad, "Imbroglio at Yale (I): Emergence of a Fraud," *Science* 210 (3 October 1980): 38–41.

105. Chris Raymond, "Allegations of Plagiarism of Scientific Manuscript Raise Concerns about 'Intellectual Theft'," *Chronicle of Higher Education* 35 (19 July 1989): A6.

106. Raymond, "Allegations of Plagiarism"; News release from *Science* (AAAS, 7 July 1989); "Notification to Readers," *Science* 245 (14 July 1989): 112; Michael Specter, "NIH Accuses Biologist of Stealing Ideas from Rival Researcher," *The Washington Post* 112 (13 July 1989): A16; Chris Raymond, "Institutes of Health Find that Purdue Professor Plagiarized Experiment," *Chronicle of Higher Education* 35 (12 July 1989): A2; Pamela S. Zurer, "NIH Panel Strips Researcher of Funding after Plagiarism Review," *Chemical & Engineering News* 67 (7 August 1989): 24–25; William J. Broad, "Question of Scientific Fakery Is Raised in Inquiry," *New York Times* (12 July 1989): A16; and Barbara J. Culliton, "NIH Sees Plagiarism in Vision Paper," *Science* 245 (14 July 1989): 120–122.

107. Culliton, "NIH Sees Plagiarism," 121; Ralph A. Bradshaw, Alice J. Adler, Frank Chytil, Nelson D. Goldberg, and Burton J. Litman, "Bridges Committee Procedures," *Science* 250 (2 November 1990): 611.

108. Raymond, "Allegations of Plagiarism," A4.

109. Quoted in Zurer, "NIH Panel Strips Researcher of Funding after Plagiarism Review," 24.

110. Culliton, "NIH Sees Plagiarism," 122; and Zurer, "NIH Panel," 24.

111. "Notification to Readers." (See n. 106 above.)

112. Broad, "Question of Scientific Fakery"; and Joseph Palca, "New Round in *Dingell v. NIH?*," *Science* 245 (28 July 1989): 349.

113. Raymond, "Allegations of Plagiarism," A6.

114. News release from *Science* (Washington, D.C.: American Association for the Advancement of Science, 7 July 1989), 2.

115. Daniel E. Koshland, Jr., "The Process of Publication," *Science* 245 (11 August 1989): 573.

116. The American Institute of Chemists' Code of Ethics, for example, specifically mentions review, noting the chemist's duty "to review the professional work of other chemists . . . fairly and in confidence." American Institute of Chemists, Code of Ethics, Section 9.

117. Articles in his journal probably never received "much, if any, external review, despite the eminence of his editorial board." Jones, "Obsession Plus Pseudo-Science," 52.

118. Agatha Christie, *The Mystery of the Blue Train* (New York: Avenel Books, 1989), 98 [originally published in 1936].

119. Kingsley Amis, *Lucky Jim* (London: Penguin Books, 1961) [originally published London: Gollancz, 1954].

120. Ibid., 14.

121. Ibid., 14.

122. Ibid., 193.

123. Ibid., 193.

124. Ibid., 193.

125. Ibid., 229.

126. Yvonne Brackbill and André E. Hellegers, "Ethics and Editors," *Hastings Center Report* 10 (February 1980): 21.

127. Ibid., 20–22. They cite Robert Veatch, ed., *Case Studies in Medical Ethics* (Cambridge: Harvard University Press, 1977), 288–289.

128. Brackbill and Hellegers, "Ethics and Editors."

129. Arnold Relman in Brackbill and Hellegers, "Ethics and Editors," 22.

130. Eleanor Singer in Brackbill and Hellegers, "Ethics and Editors," 24.

131. American Medical Association News Release, 5 July 1985.

132. Dixon, "John Maddox Offers," 13.

133. Ibid., 13.

134. "NIH's Raub on Misconduct," *The Scientist* 1 (15 December 1986): 19.

134. Marcia Angell, remarks in session on "Scientific Fraud and the Pressure to Publish," Council of Biology Editors, Annual Meeting, Raleigh, N.C., 20 May 1986.

135. Drummond Rennie, "Guarding the Guardians: A Conference on Editorial Peer Review," *Journal of the American Medical Association* 256 (7 November 1986): 2392.

136. "The Morality of Scientists," *Minerva* 25 (Winter 1987): 512.

137. Culliton, "NIH Sees Plagiarism," 121.

138. Crandall, "Peer Review," 607.

139. John Maddox, "Privacy and the Peer-Review System," *Nature* 312 (6 December 1984): 497.

CHAPTER 6: EXPOSURE

1. Angus Wilson, *Anglo-Saxon Attitudes* (New York: Viking, 1956), 270 and 271.

2. Daniel E. Koshland, Jr., "Science, Journalism, and Whistle-Blowing," *Science* 240 (29 April 1988): 585.

3. See, for example, Robert M. Anderson, Robert Perucci, Dan E. Schendel, and Leon E. Trachtman, *Divided Loyalties: Whistle-Blowing at BART* (West Lafayette: Purdue Research Foundation, 1980); Charles Peters and Taylor Branch, *Blowing the Whistle: Dissent in the Public Interest* (New York: Praeger, 1972); and Wade L. Robison, Michael S. Pritchard, and Joseph Ellin, eds., *Profits and Professions: Essays in Business and Professional Ethics* (Clifton: Humana Press, 1983).

4. Dorothy L. Sayers, *Gaudy Night* (New York: Harper & Row, 1936; New York: Avon Books, 1968).

5. Ibid., 280.

6. Ibid., 283–284.

7. Ibid., 284.

8. Wilson, *Anglo-Saxon Attitudes*.

9. Ibid., 44.

10. Ibid., 8–9.

11. Ibid., 164.

12. Ibid., 164.

13. Ibid., 262 and 272.

14. Ibid., 285.

15. Allan Mazur, "Allegations of Dishonesty in Research and Their Treatment by American Universities," *Minerva* 27 (Summer–Autumn 1989): 177–194, summarizes cases of scientific fraud and describes various nemesis and whistleblower figures.

16. Linda J. Ferguson and Janet P. Near, "Whistle-Blowing in the Lab," presented to American Criminological Society, Annual Meeting, 1989.

17. Research cited in Ferguson and Near, "Whistle-Blowing."

18. Ferguson and Near, "Whistle-Blowing," 2.

19. Based on accounts in Alison Bass, "Plagiarism Topples Leading Scientist," *Boston Globe* 234 (29 November 1988): 1 and 31; Alison Bass and Peter Gosselin, "Plagiarism First Reported to Magazine, Not Harvard," *Boston Globe* 234 (30 November 1988): 1 and 26; Alison Bass, "Scientific Community Takes Stock of System that Winks at Plagiarism," *Boston Globe* 234 (2 December 1988): 1 and 74; Morton Hunt, "Did the Penalty Fit the Crime?," *New York Times Magazine* (14 May 1989): 36–37, 69–70, 73, 84, and 98; Kim McDonald, "Noted Harvard Psychiatrist Resigns Post after Faculty Group Finds He Plagiarized," *Chronicle of Higher Education* 35 (7 December 1988): A1 and A6; Barbara J. Culliton, "Harvard Psychiatrist Resigns," *Science* 242 (2 December 1988): 1239–1240; Barbara J. Culliton, "Frazier Reinstated at McLean," *Science* 243 (17 February 1989): 889; Barbara J. Culliton, "Frazier Honored by Psychiatrists," *Science* 244 (12 May 1989): 649; and "Doctor Who Plagiarized Gets New Job," *Washington Post* (11 February 1989): A13.

20. Hunt, "Did the Penalty Fit," 69.

21. Ibid.; and Bass and Gosselin, "Plagiarism First Reported," 1.

22. Bass and Gosselin, "Plagiarism First Reported," 26.

23. Hunt, "Did the Penalty Fit," 69.

24. Bass and Gosselin, "Plagiarism First Reported," 26; Hunt, "Did the Penalty Fit," 69.

25. Mazur, "Allegations of Dishonesty," 187; Alexander Kohn, *False Prophets: Fraud and Error in Science and Medicine* (New York: Basil Blackwell, 1986): 104–110.

26. John A. Talent, "The 'Misplaced' Fossils," *Science* 246 (10 November 1989): 740–741. Also see Vishwa Jit Gupta, "The Peripatetic Fossils: Part 2," *Nature* 341 (7 September 1989): 11–12.

27. Talent, "The 'Misplaced' Fossils," 740.

28. The earlier investigations sponsored by *Nature* are described in Bernard Dixon, "Criticism Builds over *Nature* Investigation," *The Scientist* 2 (5 September 1988): 1, 4–5; and Bernard Dixon, "A Brief History of Dubious Science," *The Scientist* 2 (5 September 1988): 5.

29. Philip M. Boffey, "Two Critics of Science Revel in the Role," *New York Times* 137 (19 April 1988): C1 and C7.

30. Daniel J. Koshland, Jr., "Science, Journalism, and Whistle-Blowing," *Science* 240 (29 April 1988): 585; Michael Heylin, "Science and the Great Unwashed," *Chemical & Engineering News* 66 (11 July 1988): 3; and letter to the editor, 5 September 1988, *The Scientist*.

31. Susan Okie, "When Researchers Disagree," *Washington Post* 111 (11 April 1988): A1; Constance Holden, "Whistle-Blowers Air Cases at House Hearings," *Science* 240 (22 April 1988): 386; Daniel S. Greenberg, "Fraud Inquiry: NIH on the Capitol Griddle (continued)," *Science &*

Government Report 18 (1 May 1988): 4; Philip M. Boffey, "Nobel Winner Is Caught Up in a Dispute over Study," *New York Times* 137 (12 April 1988): C10; Barbara J. Culliton, "A Bitter Battle Over Error (II)," *Science* 241 (1 July 1988): 18; and Daniel S. Greenberg, "Lab-Scam: How Science Goes Bad," *Washington Post* 111 (25 April 1988): D3.

32. Barbara J. Culliton, "A Bitter Battle," 18.

33. Ibid., 18–20.

34. For example, Stewart is "the principal cop on the case" whose "main line of work these days is fraud-busting" [Barbara J. Culliton, "Dingell v. Baltimore," *Science* 244 (28 April 1989): 412–413]; "self-appointed fraud busters . . . their status as science's policemen uncertain" [Barbara J. Culliton, "Fraudbusters Back at NIH," *Science* 248 (29 June 1990): 1599].

35. Frank Kuznik, "Fraud Busters," *Washington Post Magazine* (14 April 1991): 24.

36. Quoted in Alexander Lindey, *Plagiarism and Originality* (New York: Harper and Brothers, 1952), 232.

37. Stephen Fay, Lewis Chester, and Magnus Linklater, *Hoax: The Inside Story of the Howard Hughes-Clifford Irving Affair* (New York: The Viking Press, 1972), 308.

38. Leon Shaskolsky Sheleff, *The Bystander: Behavior, Law, Ethics* (Lexington: Lexington Books, 1978).

39. Kuznik, "Fraud Busters," 25.

40. F. L. Jones, "Obsession Plus Pseudo-Science Equals Fraud: Sir Cyril Burt, Intelligence, and Social Mobility," *Australian and New Zealand Journal of Sociology* 16 (March 1980): 52.

41. William J. Broad and Nicholas Wade, *Betrayers of the Truth: Fraud and Deceit in the Halls of Science* (New York: Simon and Schuster, 1982), 216.

42. Kuznik, "Fraud Busters," 31.

43. Pamela S. Zurer, "Whistleblower Rejects Apology by Baltimore," *Chemical & Engineering News* 69 (13 May 1991), 7.

44. Daniel S. Greenberg, "Fraud Inquiry: Harsh Treatment for NIH on Capitol Hill," *Science & Government Report* 18 (15 April 1988): 1.

45. Boffey, "Nobel Winner Is Caught," C1, C10; Barbara J. Culliton, "Baltimore Cleared of All Fraud Charges," *Science* 243 (10 February 1989): 727; Barbara J. Culliton, "Whose Notes Are They?," *Science* 244 (19 May 1989):765; Daniel S. Greenberg, "The Hearings According to Chairman John Dingell," *Science & Government Report* 19 (15 June 1989): 1–4.

46. "The Whistle Blower on Fraud at MIT," *Boston Globe* (25 March 1991); "Scientific Fraud," *Washington Post* (28 March 1991); Philip J. Hilts, "Hero in Exposing Science Hoax Paid Dearly," *New York Times* (22 March 1991): A1 and B6.

47. American Historical Association, "Statement on Plagiarism," *Perspectives* 24 (October 1986): 8.

48. David Stipp, "Competition in Science Seems to be Spawning Cases of Bad Research," *Wall Street Journal* CCVI, No. 9 (12 July 1985): 15.

49. Hunt, "Did the Penalty Fit," 98.

50. Mark B. Roman, "When Good Scientists Turn Bad," *Discover* (April 1988): 50–58.

51. See testimony of Robert L. Sprague (pp. 183–189), letter to Robert L. Sprague from George A. Huber and letter to Robert L. Sprague from Wesley W. Posvar (pp. 200–202), reprinted in Subcommittee on Investigations and Oversight, House Committee on Science, Space, and Technology, *Maintaining the Integrity of Scientific Research*, 101st Congress, 1st session (Washington: U.S. Government Printing Office, 1990). Also see Robert L. Sprague, "I Trusted the Research System," *The Scientist* 1 (14 December 1987): 11–12.

52. Jerome Jacobstein, "I Am Not Optimistic," *The Scientist* 1 (14 December 1987): 11–12; Daniel S. Greenberg, "Part I: Chronicle of a Scientific Misconduct Case," *Science & Government Report* 16 (1 November 1986): 1–4; Daniel S. Greenberg, "Part II: Chronicle of a Scientific Misconduct Case," *Science & Government Report* 16 (15 November 1986): 1–5; Daniel S. Greenberg, "Part III: Chronicle of a Scientific Misconduct Case," *Science & Government Report* 16 (1 December 1986): 1–4.

53. Letter from Rep. Robert Roe to Secretary Otis Bowen, letter from Rep. Robert Roe to Commissioner Frank Young, and letter from Secretary Otis Bowen to Rep. Robert Roe, reprinted in *Maintaining the Integrity*, 1081–1084.

54. Bruce W. Hollis, "I Turned in My Mentor," *The Scientist* 1 (14 December 1987): 11–12.

55. Benjamin Lewin, "Fraud in Science: The Burden of Proof," *Cell* 48 (16 January 1987): 2.

56. Judy Foreman, "Probes Go Deeper in Baltimore Case," *Boston Globe* (1 April 1991): 40.

57. "The Pagliacci Murders," an untenured university professor accuses a senior male colleague of plagiarizing from "an obscure Finnish journal." A friend urges her to blow the whistle on the plagiarist, but she hesitates. She is "outrage[d]" that any scholar would "stoop so low as to plagiarize," but exposure will ruin him. When the plagiarist then murders her, the friend acts as the ultimate nemesis, avenging the crime by scaring him to death. Jules Archer, "The Pagliacci Murders," *National Forum* 67 (Winter 1987): 41–44.

58. Richard H. Blum, *Deceivers and Deceived: Observations on Confidence Men and Their Victims, Informants and Their Quarry, Political and Industrial Spies and Ordinary Citizens* (Springfield: Charles C. Thomas Publisher, 1972), 8.

59. Daniel E. Koshland, Jr., "The Process of Publication," *Science* 245 (11 August 1989): 573.

60. Philip J. Hilts, "U.S. Mental Health Agency Alleges Researcher's Work Is Fraudulent," *Boston Globe* (18 March 1987): 10.

61. Mazur, "Allegations of Dishonesty," 191.

62. Brian Evans and Bernard Waite, *IQ and Mental Testing: An Unnatural Science and Its Social History* (Atlantic Highlands: Humanities Press, 1981), 9–10, 98–107, and 159–163. Also see Nigel Hawkes, "Tracing Burt's Descent to Scientific Fraud," *Science* 205 (17 August 1979): 673.

63. Broad and Wade, *Betrayers of the Truth*, 207; and Evans and Waite, *IQ and Mental Testing*, 9.

64. Jones, "Obsession Plus Pseudo-Science," 53.

65. See Nicholas Wade, "IQ and Heredity: Suspicion of Fraud Beclouds Classic Experiment," *Science* 194 (1976): 916–919.

66. See Jones, "Obsession Plus Pseudo-Science," 53; and Broad and Wade, *Betrayers of the Truth*, 207.

67. Hawkes, "Tracing Burt's Descent." Also see L. S. Hearnshaw, *Cyril Burt, Psychologist* (London: Hodder and Stoughton, 1979).

68. Described in Barbara J. Culliton, "Dingell v. Baltimore," *Science* 244 (28 April 1989): 414; and David Baltimore, "Baltimore's Travels," *Issues in Science and Technology* 5 (Summer 1989): 48–54.

69. Larry Thompson, "Science under Fire," *Health* [supplement to the *Washington Post*] 5 (9 May 1989): 12–16; and Lisa C. Heinz, unpublished notes on meeting 2 May 1989.

70. Christopher Myers, "House Panel Renews His Criticism of the Way Universities Police Scientific Misconduct," *Chronicle of Higher Education* 36 (23 May 1990): A25; Barbara J. Culliton, "The Case of the Altered Notebooks: Part IV," *Science* 248 (18 May 1990): 809; Jeffrey Mervis, "NIH Cuts Research Funding of Scientist Under Investigation for *Cell* Paper Data," *The Scientist* 4 (11 June 1990): 9.

71. *Medicine and the Media: Ethical Problems in Biomedical Communication*, Proceedings of a symposium presented as part of the 50th anniversary celebration of the University of Rochester Medical Center, Rochester, N.Y., 9–10 October 1975, 126–129 [mimeographed transcript].

72. Hilts, "Hero in Exposing Science Hoax," B6.

73. Joseph Palca, "Justice Department Joins Whistle-Blower Suit," *Science* 249 (17 August 1990): 734; Colleen Cordes, "U.S. Enters Lawsuit Accusing Scientist, Institutions of Fraud," *Chronicle of Higher Education* 37 (5 September 1990): A1 and A24; and Pamela S. Zurer, "Scientific Misconduct: U.S. Joins Research Fraud Suit," *Chemical & Engineering News* 68 (20 August 1990): 6–7.

74. Kuznik, "Fraud Busters," 32.

CHAPTER 7: ACTION

1. George Savage, *Forgeries, Fakes, and Reproductions: A Handbook for the Art Dealer and Collector* (New York: Frederick A. Praeger, 1963), 20.

2. Bernard Ashmole, *Forgeries of Ancient Sculpture, Creation and Detection*, the first J. L. Myres Memorial Lecture, delivered in New College, Oxford, on 9 May 1961 (Oxford, England: Holywell Press, 1962), 3.

3. Dennis Rosen, "The Jilting of Athene," *New Scientist* 39 (5 September 1968): 500.

4. Thomas Mallon, *Stolen Words: Forays into the Origins and Ravages of Plagiarism* (New York: Ticknor and Fields, 1989), 135.

5. C. P. Snow, *The Affair* (New York: Charles Scribner's Sons, 1960).

6. Ibid., 7.

7. Ibid., 15.

8. Ibid., 21.

9. Ibid., 76.

10. Ibid., 89.

11. Ibid., 91.

12. Ibid., 31.

13. Ibid., 123.

14. Ibid., 112.

15. Ibid.

16. Ibid., 159.

17. Ibid.

18. Ibid., 186.

19. Ibid., 245.

20. Ibid., 245.

21. Ibid., 332.

22. Constance Holden, "NIH Moves to Debar Cholesterol Researcher," *Science* 237 (14 August 1987): 718–719, describes allegations made against Charles J. Glueck, who acknowledged that flaws existed in a paper he had published but maintained that any errors were inadvertent, random, and unintentional.

23. Holden, "NIH Moves to Debar," 718.

24. Mallon, *Stolen Words*, 171.

25. Ibid., 180–182.

26. Leslie Roberts, "Caltech Deals with Fraud Allegations," *Science* 251 (1 March 1991): 1014.

27. Benjamin Lewin, "Fraud in Science: The Burden of Proof," *Cell* 48 (16 January 1987): 1–2.

28. Joseph Palca, "Investigations into NIH Fraud Allegations End with Suicide," *Nature* 325 (19 February 1987): 652.

29. Lewin, "Fraud in Science."

30. Council of Biology Editors, *Economics of Scientific Journals* (Bethesda: CBE, 1982).

31. William J. Broad, "Harvard Delays in Reporting Fraud," *Science* 215 (29 January 1982): 478. Broad said that "Harvard University took public action 6 months after the admitted falsification."

32. Ron Dagani, "*Nature* Publishes Paper It Finds Unbelievable," *Chemical & Engineering News* 66 (4 July 1988): 5; Ron Dagani, "*Nature* Refutes Research It Had Published Earlier," *Chemical & Engineering News* 66 (1 August 1988): 6; Colleen Cordes, "Journal Discredits Research Results It Published," *Chronicle of Higher Education* 34 (3 August 1988): A1 and A10; Bernard Dixon, "Criticism Builds over *Nature* Investigation," *The Scientist* 2 (5 September 1988): 1, 4–5; Robert Pool, "Unbelievable Results Spark a Controversy," *Science* 241 (22 July 1988): 407; and Robert Pool, "More Squabbling Over Unbelievable Result," *Science* 241 (5 August 1988): 658.

33. Dagani, "*Nature* Publishes," 5; Dagani, "*Nature* Refutes," 6; Pool, "Unbelievable Results," 407; Cordes, "Journal Discredits Research," A1 and A10; Pool, "More Squabbling," 658. E. Davenas, F. Beauvais, J. Amara, M. Oberbaum, B. Robinzon, A. Miadonna, A. Tedeschi, B. Pomeranz, P. Fortner, P. Belon, J. Sainte-Laudy, B. Poitevin, and J. Benveniste, "Human Basophil Degranulation Triggered by Dilute Serum against IgE," *Nature* 333 (30 June 1988): 816–818; John Maddox, "When to Believe the Unbelievable," *Nature* 333 (30 June 1988): 787; John Maddox, James Randi, and Walter W. Stewart, "'High-dilution' Experiments a Delusion," *Nature* 334 (28 July 1988): 287–290; and Jacques Benveniste, "Dr. Jacques Benveniste Replies," *Nature* 334 (28 July 1988): 291.

At the end of the 30 June 1988 article by E. Davenas et al., *Nature* had added an "Editorial reservation" (p. 818) which read: "Readers of this article may share the incredulity of the many referees who have commented on several versions of it during the past several months." Then, after describing the phenomenon, it added that "there is no physical basis for such an activity."

34. Dagani, "*Nature* Publishes," 5.

35. Ibid., 6; Cordes, "Journal Discredits Research."

36. Dixon, "Criticism Builds," 1.

37. Dagani, "*Nature* Refutes," 6.

38. Pool, "More Squabbling," 658.

39. Dixon, "Criticism Builds," 4.

40. Ibid., 4.

41. Bernard Dixon, "A Brief History of Dubious Science," *The Scientist* 2 (5 September 1988): 5.

42. Eugene Garfield, "Contrary to *Nature?*," *The Scientist* 2 (5 September 1988): 12.

43. Pool, "More Squabbling," 658.

44. Arnold S. Relman, "Relman on Maddox," *Science* 242 (21 October 1988): 348.

45. Ibid., 348.

46. David Dickson, "Benveniste Criticism Is Diluted," *Science* 245 (21 July 1989): 248.

47. Stephen Lock, *A Difficult Balance: Editorial Peer Review in Medicine* (Philadelphia: ISI Press, 1986), 54, citing William Broad and Nicholas Wade, *Betrayers of the Truth*, 94–95.

48. William Booth, "Voodoo Science," *Science* 240 (15 April 1988): 275.

49. Wade Davis, "Zombification," *Science* 240 (24 June 1988): 1716.

50. Wally Herbert, "Did Peary Reach the Pole?," *National Geographic* 174 (September 1988): 412.

51. Lewin, "Fraud in Science," 2.

52. Hope Werness, "Han van Meegeran fecit," in *The Forger's Art: Forgery and the Philosophy of Art*, ed. Denis Dutton (Berkeley, Los Angeles, London: University of California Press, 1983), 50.

53. Arthur Sherbo, "The Uses and Abuses of Internal Evidence," in *Evidence for Authorship: Essays on Problems of Attribution*, ed. David V. Erdman and Ephim G. Fogel (Ithaca: Cornell University Press, 1966), 6–24.

54. Arthur Pierce Middleton and Douglass A. Adair, "The Mystery of the Horn Papers," in Erdman and Fogel, *Evidence for Authorship*.

55. Ibid., 384.

56. Erdman and Fogel, *Evidence for Authorship*, 377.

57. Nigel Hawkes, "Tracing Burt's Descent to Scientific Fraud," *Science* 205 (17 August 1979): 673. Also see Broad and Wade, *Betrayers of the Truth*, 206.

58. L. S. Hearnshaw, *Cyril Burt, Psychologist* (London: Hodder and Stoughton, 1979).

59. William M. Calder and David A. Traill, eds., *Myth, Scandal, and History: The Heinrich Schliemann Controversy and a First Edition of the Mycenaean Diary* (Detroit: Wayne State University Press, 1986).

60. Ibid., 9.

61. F. L. Jones, "Obsession Plus Pseudo-Science Equals Fraud: Sir Cyril Burt, Intelligence, and Social Mobility," *Australian and New Zealand Journal of Sociology* 16 (March 1980): 48.

62. Ibid., 48.

63. Ashmole, *Forgeries of Ancient Sculpture*, 3.

64. Ibid., 3.

65. Barbara J. Culliton, "NIH Sees Plagiarism in Vision Paper," *Science* 245 (14 July 1989): 121. Also see Pamela S. Zurer, "NIH Panel Strips Researcher of Funding after Plagiarism Review," *Chemical & Engineering News* 67 (7 August 1989): 24–25.

66. Zurer, "NIH Panel," 25; Culliton, "NIH Sees Plagiarism," 122.

67. Barbara J. Culliton, "The Dingell Probe Finally Goes Public," *Science* 244 (12 May 1989): 643–646; Eliot Marshall, "Secret Service Probes Lab Notebooks," *Science* 244 (12 May 1989): 644; and David P. Hamilton, "NIH Finds Fraud in *Cell* Paper," *Science* 251 (29 March 1991): 1552–1554.

68. Testimony by John W. Hargett and Larry F. Stewart, U.S. Congress, House Committee on Energy and Commerce, Subcommittee on Oversight and Investigation, "Scientific Fraud," 101st Congress, 2d session, 14 May 1990. Also see Pamela S. Zurer, "Scientific Misconduct: Criminal Inquiry on Scientist Urged," *Chemical & Engineering News* 68 (21 May 1990): 6; David P. Hamilton, "Verdict in Sight in the 'Baltimore Case,'" *Science* 251 (8 March 1991): 1168–1172.

69. Quoted by Thomas Hoving in John McPhee, "A Roomful of Hovings," *A Roomful of Hovings and Other Profiles* (New York: Farrar, Straus and Giroux, 1968), 23.

70. For example, M. B. Mittleman, "Fraudulent Science," *Science* 239 (8 January 1988): 127; and Lee Loevinger, "Anonymous Charges," *Science* 239 (19 February 1988): 847.

71. Zurer, "NIH Panel," 24.

72. Peter Novick, *That Noble Dream: The 'Objectivity Question' and the American Historical Profession* (New York: Cambridge University Press, 1988).

73. Ibid., 618.

74. Barbara Mishkin, "Responding to Scientific Misconduct: Due Process and Prevention," *Journal of the American Medical Association* 260 (7 October 1988): 1932–1936.

CHAPTER 8: RESOLUTION

1. Thomas Mallon, *Stolen Words: Forays into the Origins and Ravages of Plagiarism* (New York: Ticknor and Fields, 1989), 131.

2. R. T. Campbell, *Unholy Dying: A Detective Story* (London: John Westhouse Ltd., 1945; New York: Dover Publications, Inc., 1985), 11. "R. T. Campbell" was the pseudonym of Ruthven Campbell Todd (1914–1978), British art critic, poet, and novelist. Perhaps because of his own university connections, Campbell offers a disclaimer ("Apologia") at the beginning of *Unholy Dying*: "An apology is due to all geneticists, none of whom, I am sure, would behave in the least like the characters in this book." (p. vi)

3. Campbell, *Unholy Dying*, 84.

4. Quoted in Michael Heylin, "AAAS Probes Penalty for Scientific Fraud," *Chemical & Engineering News* 68 (5 March 1990): 16.

5. Edward Sagarin, *Deviants and Deviance: An Introduction to the Study of Disvalued People and Behavior* (New York: Praeger Publishers, 1975), 14. He cites Erving Goffman's term "remedial work."

6. John Maddox, "How to Say Sorry Graciously," *Nature* 338 (2 March 1989): 13.

7. Acknowledging that he "did not give sufficient credit to some of the writing which had been helpful" in preparing a profile of James Agee, John Hersey stated: "I very much regret any discourtesy on my part, or any distress caused . . . by these omissions." [*The New Yorker* (1 August 1988)], quoted by M. R. Montgomery in the *Boston Globe* (August 1988).

8. Pamela S. Zurer, "Scientific Whistleblower Vindicated," *Chemical & Engineering News* 69 (8 April 1991): 35–36, 40; Judy Foreman, "Baltimore Seems to be Moving to Clear His Name," *Boston Globe* (24 April 1991): 13; David P. Hamilton, "NIH Finds Fraud in *Cell* Paper," *Science* 251 (29 March 1991): 1552–1554; David P. Hamilton, "Baltimore Throws in the Towel," *Science* 251 (10 May 1991): 768–770; Philip J. Hilts, "'I Am Innocent,' Embattled Biologist Says," *New York Times* (4 June 1991): C1 and C10; and David P. Hamilton, "Baltimore Case—In Brief," *Science* 253 (5 July 1991): 24–25.

9. "David Baltimore's Mea Culpa," *Science* 252 (10 May 1991): 769–770. See also Pamela S. Zurer, "Whistleblower Rejects Apology by Baltimore," *Chemical & Engineering News* 69 (13 May 1991): 7.

10. Melvin E. Jahn and Daniel J. Woolf, eds., *The Lying Stones of Dr. Johann Bartholomew Adam Beringer, being his Lithographiae Wirceburgensis* (Berkeley and Los Angeles: University of California Press, 1963).

11. Leonard B. Meyer, "Forgery and the Anthropology of Art," in *The Forger's Art: Forgery and the Philosophy of Art*, ed. Denis Dutton (Berkeley, Los Angeles, London: University of California Press, 1983), 89, citing Emily Genauer.

12. Mallon, *Stolen Words*, 123.

13. Ian Haywood, *Faking It: Art and the Politics of Forgery* (London: The Harvester Press, 1987), 133.

14. Dennis Rosen, "The Jilting of Athene," *New Scientist* 39 (5 September 1968): 500, calls "a simple published retraction . . . the best way of dealing with scientific errors."

15. Graham DuShane, Konrad B. Krausfopf, Edwin M. Lerner, Philip M. Morse, H. Burr Steinbach, William L. Straus, Jr., and Edward L. Tatum, "An Unfortunate Event," *Science* 134 (September 1961): 945.

16. Ibid., 946.

17. Ibid., 945.

18. Efraim Racker, "The Warburg Effect: Two Years Later," *Science*

222 (21 October 1983): 232. Also see Jeffrey L. Fox, "Theory Explaining Cancer Partly Retracted," *Chemical & Engineering News* 59 (7 September 1981): 35–36; and Nicholas Wade, "The Rise and Fall of a Scientific Superstar," *New Scientist* 91 (24 September 1981): 781–782.

19. Kathleen Case, speaking in a session on "Retractions and Corrections in Primary and Secondary Sources," Council of Biology Editors, Raleigh, N.C., 20 May 1986.

20. Figures given in Allan Mazur, "Allegations of Dishonesty in Research and Their Treatment by American Universities," *Minerva* 27 (Summer–Autumn 1989): 185. Listed in: *Diabetes* 29 (1980): 672; *American Journal of Medicine* 69 (1980): 17.

21. As reprinted in *Annals of Internal Medicine* 108 (1988): 304.

22. "Scientific Fraud: In Bristol Now," *Nature* 294 (10 December 1981): 509.

23. Efraim Racker and Mark Spector, "Warburg Effect Revisited," *Science* 213 (18 September 1981): 1313–1314, was commenting on Efraim Racker and Mark Spector, "Warburg Effect Revisited: Merger of Biochemistry and Molecular Biology," *Science* 213 (17 July 1981): 303–307.

24. Racker and Spector, "Warburg Effect Revisited" (September).

25. Ibid.

26. Efraim Racker, "A View of Misconduct in Science," *Nature* 339 (11 May 1989): 91–93.

27. Claudio Milanese, Neil E. Richardson, and Ellis L. Reinberg, "Retraction of Data," *Science* 234 (28 November 1986): 1056.

28. Ibid.

29. Constance Holden, "NIH Moves to Debar Cholesterol Researcher," *Science* 237 (14 August 1987): 718–719.

30. E. Frederick Wheelock, "Plagiarism and Freedom of Information Laws," *Lancet* (12 April 1980): 826.

31. Mazur, "Allegations of Dishonesty," 190.

32. Research communications were published by Ronald Breslow and Monica Mehta in *Journal of American Chemical Society* 108 (1986): 2485 and 6417. These were first retracted by Ronald Breslow in *Chemical & Engineering News* 64 (8 December 1986). Also see Ronald Breslow, "Scientific Fraud," *Chemical & Engineering News* 65 (11 May 1987): 2.

33. Monica P. Mehta, "Scientific Fraud," *Chemical & Engineering News* 65 (11 May 1987): 2.

34. Formal report of the Columbia University investigating committee is reprinted in U.S. Congress, House Committee on Science, Space and Technology, Subcommittee on Investigations and Oversight, *Maintaining the Integrity of Scientific Research*, 101st Congress, 1st session (Washington, D.C.: U.S. Government Printing Office, 1990), 1035–1065.

35. "The Morality of Scientists: I—Procedures for Responding to Charges of Unethical Research Practices: University of California, San

Diego, School of Medicine," and "The Morality of Scientists: II—Report of the Faculty *ad hoc* Committee to Investigate Research Fraud," *Minerva* 25 (Winter 1987): 502–512. See also Eliot Marshall, "San Diego's Tough Stand on Research Fraud," *Science* 234 (31 October 1986): 534–535; Rex Dalton, "Fraudulent Papers Stain Co-Authors," *The Scientist* 1 (18 May 1987): 1, 8; Robert L. Engler, James W. Covell, Paul J. Friedman, Philip S. Kitcher, and Richard M. Peters, "Misrepresentation and Responsibility in Medical Research," *New England Journal of Medicine* 317 (26 November 1987): 1383–1389; and Paul J. Friedman, "Research Ethics, Due Process, and Common Sense," *Journal of the American Medical Association* 260 (7 October 1988): 1937–1938.

36. "The Morality of Scientists," 504.

37. Ibid., 506.

38. Anthony Bailey, "A Young Man on Horseback," *The New Yorker* 66 (5 March 1990): 50.

39. "The Morality of Scientists," 507. "Statement Concerning Research Fraud," *Cardiovascular Research* 21 (1987): 240, is an example of an announcement that appeared in one journal.

40. Friedman, remarks to the American Association for the Advancement of Science, Annual Meeting, Chicago, February 1986 [unpaginated manuscript]; also see Friedman, "Research Ethics."

41. Marshall, "San Diego's Tough Stand," 535, quoting the managing editor of *Radiology*.

42. Rex Dalton, "Journals Slow to Retract Slutsky Research Errors," *The Scientist* 2 (7 March 1988): 1 and 4, reporting that eleven journals had published retraction notices.

43. Rosen makes this point in "The Jilting of Athene," 500.

44. Ibid., 500.

45. Ibid., 500.

46. Dalton, "Journals Slow," 4.

47. Ibid., 4.

48. Ibid., 4.

49. Ibid., 4.

50. Rex Dalton, "Panel Backs Journal Retractions," *The Scientist* 2 (7 March 1988): 4.

51. Kathleen Case, "Retractions and Corrections in Primary and Secondary Sources," Council of Biology Editors, Annual Meeting, Raleigh, N.C., 20 May 1986, commenting on the August 1983 issue of the journal.

52. Employee of Chemical Abstracts, speaking at "Retractions and Corrections in Primary and Secondary Sources," Council of Biology Editors, Annual Meeting, Raleigh, N.C., 20 May 1986.

53. For example, Donald A. B. Lindberg, "Retraction of Research Findings," *Science* 235 (13 March 1987): 1308.

54. Robin Shivers, discussant in "Retractions and Corrections in Pri-

mary and Secondary Sources," Council of Biology Editors, Annual Meeting, Raleigh, N.C., 20 May 1986.

55. Alison Bass, "Scientific Community Takes Stock of System that Winks at Plagiarism," *Boston Globe* 234 (2 December 1988): 1 and 74.

56. MIT, *Policies and Procedures*, Section 3.51, p. 4.

57. Jeffrey Mervis, "Misconduct Office Vows to Keep Actions Secret," *The Scientist* 4 (19 March 1990): 1, 8–10.

58. Ibid., 9.

59. Paul Friedman, quoted in Mervis, "Misconduct Office Vows," 10.

60. Patricia Woolf, quoted in Mervis, "Misconduct Office Vows," 10.

61. Friedman, remarks to AAAS.

62. A. J. Hostetler, "Fear of Suits Blocks Retractions," *The Scientist* 1 (19 October 1987): 1–2.

63. Ibid., 1.

64. Rex Dalton, "Panel Backs Journal Retractions," *The Scientist* 2 (7 March 1988): 4.

65. Barbara Mishkin and Arthur F. Schwartz, "Professional Journals: The Responsibility to Correct the Record," *Professional Ethics Report* 1 (Fall 1988): 1–2.

66. National Science Foundation, Office of Inspector General, Circular OIG 90–2, 28 September 1990.

67. Barbara J. Culliton, "NIH Misconduct Probes Draw Legal Complaints," *Science* 249 (20 July 1990): 240.

68. Introduced by Representative Henry A. Waxman (D-California), September 1988.

69. Barbara J. Culliton, "NIH Sees Plagiarism in Vision Paper," *Science* 245 (14 July 1989): 120–122.

70. William J. Broad, "Question of Scientific Fakery Is Raised in Inquiry," *New York Times* (12 July 1989): A16.

71. Mervis, "Misconduct Office Vows."

72. Colin Norman, "Prosecution Urged in Fraud Case," *Science* 236 (29 May 1987): 1057.

73. David L. Wheeler, "Experts Weigh Federal Prosecution of Scholars Who Fabricate Results," *Chronicle of Higher Education* 35 (1 March 1989): A5.

74. Constance Holden, "Fraud Reimbursement," *Science* 237 (25 September 1987): 1563.

75. Irving Warshawsky, "Research Fraud," *Chemical & Engineering News* 66 (27 June 1988): 66–67.

76. Ibid., 67.

77. Barbara J. Culliton, "The Sloan-Kettering Affair (II): An Uneasy Resolution," *Science* 184 (14 June 1974): 1155 and 1157.

78. Morton Hunt, "Did the Penalty Fit the Crime?," *New York Times Magazine* (14 May 1989): 36–37, 69–70, 73, 84, and 98.

79. Stanley L. Lee, "Fraud in Medical Research," *Journal of the American Medical Association* 251 (3 January 1984): 215.

80. "Massachusetts v. Alsabti," *Science* 245 (8 September 1989): 1046.

81. Marjorie Sun, "NIH Developing Policy on Misconduct," *Science* 216 (14 May 1982): 711.

82. Rosen, "The Jilting of Athene," 500.

83. Philip Abelson proposes that authors caught in a double submission should have their future articles banned from consideration for three years. Philip H. Abelson, "Excessive Zeal to Publish," *Science* 218 (1982): 953. Cited in Stephen Lock, *A Difficult Balance: Editorial Peer Review in Medicine* (Philadelphia: ISI Press, 1986), 45.

84. Constance Holden, "Fraud Reimbursement," *Science* 237 (25 September 1987): 1563.

CHAPTER 9: ON THE HORIZON

1. Angus Wilson, *Anglo-Saxon Attitudes* (New York: The Viking Press, 1956), 8–9.

2. Martin F. Shapiro and Robert P. Charrow, "The Role of Data Audits in Detecting Scientific Misconduct," *Journal of the American Medical Association* 261 (5 May 1989): 2505–2511.

3. The president of the Institute of Medicine once suggested that editors request the original data for every two hundredth paper submitted. David L. Wheeler, "Researchers Need to Shoulder Responsibility for Detecting Misconduct, Scientist Warns," *Chronicle of Higher Education* 33 (25 February 1987): 13.

4. Drummond Rennie, "How Much Fraud? Let's Do an Experimental Audit," *AAAS Observer*, No. 3 (6 January 1989): 4; Drummond Rennie, "Editors and Auditors," *Journal of the American Medical Association* 261 (5 May 1989): 2543–2545.

5. American Association for the Advancement of Science-American Bar Association (AAAS-ABA), *Project on Scientific Fraud and Misconduct, Report on Workshop Number Three* (Washington, D.C.: AAAS, 1989). Also see Ellen K. Coughlin, "Concerns about Fraud, Editorial Bias Prompt Scrutiny of Journal Practices," *Chronicle of Higher Education* 35 (15 February 1989): A4–A7; Barbara J. Culliton, "Random Audit of Papers Proposed," *Science* 242 (4 November 1988): 657–658; Jeffrey Mervis, "Random Audits of Raw Data?," *The Scientist* 2 (28 November 1988): 1–3; Rennie, "How Much Fraud?," 4; Rennie, "Editors and Auditors," 2543–2545; and Shapiro and Charrow, "The Role of Data Audits," 2505–2511.

6. Culliton, "Random Audit," 657.

7. John Bailar, quoted in Culliton, "Random Audit," 657.

8. Bernard D. Davis, "Fraud: Why Auditing Laboratory Records Is a Bad Idea," *The Scientist* 3 (20 February 1989): 9.

9. Philip M. Boffey, "Major Study Points to Faulty Research at Two Universities," *New York Times*, 22 April 1986: C1; Daniel S. Greenberg, "Part II: The Fraud Study That's Too Hot to Publish," *Science & Government Report* 16 (1 April 1986).

10. Quoted in Kim A. McDonald, "Chemists Whose Cold-Fusion Claims Created Furor Ignite Another Controversy at the University of Utah," *Chronicle of Higher Education* 36 (6 June 1990): A8.

11. AAAS-ABA, *Report on Workshop Number Three*, 24–25, 181–186.

12. Ibid., 26; also see Nina Bang-Jensen, "Libel and Scientific Publication," in AAAS-ABA, *Report on Workshop Number Three*, 195–202.

13. Cathy Sears, "Supreme Court Ruling Could Stifle Open Debate in Journals," *The Scientist* 4 (1 October 1990): 1, 7 and 10.

14. "The Morality of Scientists: I—Procedures for Responding to Charges of Unethical Research Practices: University of California, San Diego, School of Medicine," *Minerva* 25 (Winter 1987): 502–503; and "The Morality of Scientists: II—Report of the Faculty *ad hoc* Committee to Investigate Research Fraud,"*Minerva* 25 (Winter 1987): 506.

15. AAAS-ABA, *Report on Workshop Number Three*.

16. Ibid., 17.

17. Ibid., 17, 99–110. Jonathan Weissler added in the same discussion that "it is unlikely that any regulation can be designed which will totally remove the potential for professional retribution if the whistleblower is a relatively junior individual" or is the employee of the accused.

18. Office of Technology Assessment, *Federal Scientific and Technical Information in an Electronic Age: Opportunities and Challenges* (Washington, D.C.: U.S. Government Printing Office, 1989).

19. Kathleen Carley and Kira Wendt, "Electronic Mail and Scientific Communication: A Study of the Soar Extended Research Group," *Knowledge: Creation, Diffusion, Utilization* 12 (June 1991): 406–440.

20. F. Wilfred Lancaster, "Electronic Publishing: Its Impact on the Distribution of Information," *National Forum* 63 (Spring 1984): 3–5; and Irving L. Horowitz, *Communicating Ideas: The Crisis of Publishing in a Post-Industrial Society* (New York: Oxford University Press, 1986).

21. Paul Gleason, "PSP Division Annual Meeting," *SSP Letter* 12 (March–April 1990): 2.

22. Leslie Roberts, "The Rush to Publish," *Science* 251 (18 January 1991): 260–263.

23. Ibid., 260.

24. Robert M. Hayes, "Desktop Publishing: Problems of Control," *Scholarly Publishing* 21 (January 1990): 121.

25. Paul J. Anderson, quoted in the "Footnotes" column, *Chronicle of Higher Education* 38 (22 May 1991): A5. Also see Daniel Sheridan, "The

Trouble with Harry: High Technology Can Now Alter a Moving Video Image," *Columbia Journalism Review* (January–February 1990): 4 and 6.

26. Nelson Goodman, "Art and Authenticity," in *The Forger's Art: Forgery and the Philosophy of Art*, ed. Denis Dutton (Berkeley, Los Angeles, London: University of California Press, 1983), 103–104.

27. Dennis Drabelle, "Copyright and Its Constituencies: Reconciling the Interests of Scholars, Publishers, and Librarians," *Scholarly Communication*, No. 3 (Winter 1986): 4.

28. Robert P. Benko, *Protecting Intellectual Property Rights: Issues and Controversies* (Washington, D.C.: American Enterprise Institute for Public Policy Research, 1987); Copyright Office, Library of Congress, *The United States Joins the Berne Union*, Circular 93a (Washington, D.C.: U.S. Government Printing Office, 1989); Office of Technology Assessment, U.S. Congress, *Intellectual Property Rights in an Age of Electronics and Information* (Washington, D.C.: U.S. Government Printing Office, 1986).

29. For discussion of several of these cases, see U.S. Congress, House Committee on Government Operations, *Are Scientific Misconduct and Conflicts of Interest Hazardous to Our Health?*, House Report No. 101–688 (Washington, D.C.: U.S. Government Printing Office, 1990).

30. Jeffrey Mervis, "Bitter Suit over Research Work Asks 'Who Deserves the Credit?,'" *The Scientist* 3 (17 April 1989): 1, 4–5.

31. Ibid., 4.

32. National Academy of Sciences-National Academy of Engineering (NAS-NAE), *Information Technology and the Conduct of Research: The User's View* (Washington, D.C.: National Academy of Sciences, 1989), 28, 50.

33. Statement made in a 9 May 1989 hearing before the House Committee on Energy and Commerce, Subcommittee on Oversight and Investigations, U.S. Congress. Also see Barbara J. Culliton, "Whose Notes Are They?," *Science* 244 (19 May 1989): 765.

34. Office of Technology Assessment, *Intellectual Property Rights*.

35. This paragraph benefits from discussions with a number of experts on intellectual property. Brian Kahin's work on the topic is notable because he takes a systems approach to the traditional intellectual property issues.

36. William Y. Arms, "Electronic Publishing and the Academic Library," Society for Scholarly Publishing, 11th Annual Meeting, Washington, D.C., 31 May 1989.

37. Office of Technology Assessment, *Intellectual Property Rights*, 75.

38. David Ladd, "Securing the Future of Copyright: A Humanist Endeavor," remarks before the International Publishers Association, Mexico City, 13 March 1984.

39. David H. Johnson, "Science-Policy Focus: Can Scientific Miscon-

duct Be Stopped Without New Legislation?," *CSSP News* 5 (November 1990): 5.

40. David L. Bazelon, *Questioning Authority: Justice and Criminal Law* (New York: Alfred A. Knopf, 1988), 8.

41. Ibid., 8.

42. Alexander Lindey, *Plagiarism and Originality* (New York: Harper and Brothers, 1952), 239. Lindey [p. 242] points out that it should make no difference whether the source is dead or alive; "What, if any, is the relevance and impact of such factors as the first gatherer's calling, the borrower's purpose, the ratio between the quantity taken and the total source, and the ratio between the quantity taken and the borrower's whole work?"

43. Thomas Mallon, *Stolen Words: Forays into the Origins and Ravages of Plagiarism* (New York: Ticknor and Fields, 1989), 123.

44. Walter Redfern, *Clichés and Coinages* (London: Basil Blackwell, 1989), 66.

45. Lock emphasizes the importance of balancing free expression with expert review. As part of the free press, journals have a general social obligation to provide a forum open to all points of view; as part of the scientific community, they have an obligation to publish only articles that represent the highest standards of their discipline and that do not lead readers astray. Stephen Lock, *A Difficult Balance: Editorial Peer Review in Medicine* (Philadelphia, Pennsylvania: ISI Press, 1986).

46. Lindey, *Plagiarism and Originality*, 231.

Selected Bibliography

The general news media, especially the nationally oriented newspapers like the *New York Times* and the *Washington Post*, have followed alleged cases of research fraud with increasing fervor, but by far the most complete reporting has appeared in the specialized science and academic news publications, notably *Chemical & Engineering News*, the *Chronicle of Higher Education*, *Nature*, *Science & Government Report*, and *The Scientist*. Articles from these magazines, from the mid-1970s to present, have served as sources for this book, although I have relied on them only insofar as information was widely reported, with what appeared to be corroborative interviews with the principals, and was not later disputed or rejected.

Within each specialized publication, certain writers stand out. Daniel S. Greenberg's reports in *Science & Government Report* continually emphasize the subtle political implications of the cases. The work of Pamela S. Zurer and Ron Dagani in *Chemical & Engineering News* exemplify the type of reporting that lets the facts speak—or shout—on their own.

The news section of *Science* has consistently allocated considerable news space to coverage of research fraud. William J. Broad and Nicholas Wade began that tradition when they were on the *Science* news staff and continued their interest in the topic when they later went to the *New York Times*. *Science* articles tend to include explanations of the scientific aspects of cases, as well as the ins and outs of the political and legal action. The fullest reports are those by Barbara J. Culliton, whose articles from 1974 to 1990 reflected the scientific community's own reactions to the events. Other reports in *Science* include those by William Booth, Gregory Byrne, David Dickson, David P. Hamilton, Constance Holden, Roger Lewin, Eliot Marshall, and Robert Pool. Joseph Palca started following the issue at

Nature and Leslie Roberts at *BioScience*; both continued to report on the issue when they joined the *Science* staff.

The *Chronicle of Higher Education* has traced the events in terms of how federal actions impact universities. Colleen Cordes, Ellen K. Coughlin, Kim McDonald, and David L. Wheeler have written on the topic since the mid-1980s, followed more recently by Debra Blum, Christopher Myers, and Chris Raymond. Other weekly publications that pay close attention to research fraud are *The Scientist* and *Nature*, where writers G. Christopher Anderson, Rex Dalton, Bernard Dixon, and Jeffrey Mervis have added to the record.

Perhaps because of the Boston-area connections of many major stories, some *Boston Globe* reporters, such as Alison Bass, Judy Foreman, Peter G. Gosselin, and Richard Saltus, have written extensive pieces on scientific fraud. Philip J. Hilts has written for the *Globe* as well as in the *Washington Post*, where Susan Okie and Larry Thompson have also provided comprehensive analyses of Washington area stories. The *New York Times* reports by Philip Boffey and William Broad tend to pull no punches and often round off coverage with focus on the personalities caught up in cases.

Commentary on the subject—in journal articles, editorials, and letters to the editor—forms another source of insight, but solid, research-based analyses have been noticeably lacking from the literature. I list some of the most important works in the bibliography below.

Other useful sources of information include: congressional hearings and reports, including the official transcripts from those hearings and my own notes and recordings, official reports from federal agencies, and reports from scientific organizations. The book also draws from confidential interviews with people who have been involved in circumstances similar to those discussed, and from my own experiences as a journal author, referee, editor, and editorial board member. The conclusions and recommendations are my own, but I have tried to reflect accurately the *range* of concerns expressed by those who have accused, those who have been accused, and those who have participated in difficult investigations. The intent is not to embarrass, applaud, or condemn but to draw lessons we all need to learn.

Several decades from now, historians will eventually analyze the voluminous confidential files being compiled by employers, research sponsors, government agencies, and the justice system as cases are being investigated. At present, most records remain closed. Bearing in mind therefore that the complete stories are not yet always known, I have attempted to be sensitive to competing claims, when participants either still plead innocence (or have refused to acknowledge guilt publicly) or still harbor beliefs in another's guilt or innocence.

We have much yet to learn. It is my hope that this book will serve as a starting point for that effort and for future research on the entire process and history of modern scientific communication systems.

BOOKS

Albury, Randall. *The Politics of Objectivity.* Victoria, Australia: Deakin University Press, 1983.

Arnau, Frank. *The Art of the Faker: 3,000 Years of Deception.* Boston: Little, Brown and Company, 1961.

Ashmole, Bernard. *Forgeries of Ancient Sculpture, Creation and Detection.* The first J. L. Myres Memorial Lecture, delivered in New College, Oxford, on 9 May 1961. Oxford, England: Holywell Press, 1962.

Barnes, Barry. *About Science.* London: Basil Blackwell, 1985.

Barnes, J. A. *Who Should Know What? Social Science, Privacy and Ethics.* Cambridge: Cambridge University Press, 1979.

Barzun, Jacques. *On Writing, Editing, and Publishing: Essays Explicative and Hortatory,* 2d ed. Chicago: University of Chicago Press, 1986.

Beauchamp, Tom L., and Norman E. Bowie, eds. *Ethical Theory and Business.* Englewood Cliffs: Prentice-Hall, Inc., 1979.

Beauchamp, Tom L., Ruth R. Faden, R. Jay Wallace, Jr., and LeRoy Walters, eds. *Ethical Issues in Social Science Research.* Baltimore: Johns Hopkins University Press, 1982.

Bennington, Harold. *The Detection of Frauds.* Chicago: LaSalle Extension University, 1952.

Berg, Kåre, and Knut Erik Tranøy, eds. *Research Ethics.* New York: Alan R. Liss, Inc., 1983.

Bezanson, Randall P., Gilbert Cranberg, and John Soloski. *Libel Law and the Press: Myth and Reality.* New York: The Free Press, 1987.

Blackman, George. *On Bartholow and Pro's "Liberal Use" of Prize Essays; or Prize-Essaying Made Easy and Taught in a Single Lesson.* Cincinnati: Wrightson & Co., 1868.

Blinderman, Charles. *The Piltdown Inquest.* Buffalo: Prometheus Books, 1986.

Blum, Richard H. *Deceivers and Deceived; Observations on Confidence Men and Their Victims, Informants and Their Quarry, Political and Industrial Spies and Ordinary Citizens.* Springfield: Charles C. Thomas, 1972.

Blumenthal, Walter Hart. *False Literary Attributions, Works Not Written by Their Supposed Authors, or Doubtfully Ascribed.* Philadelphia: George S. MacManus Company, 1965.

Broad, William J., and Nicholas Wade. *Betrayers of the Truth: Fraud and Deceit in the Halls of Science.* New York: Simon and Schuster, 1982.

Calder, William M., and David A. Traill, eds. *Myth, Scandal, and History: The Heinrich Schliemann Controversy and a First Edition of the Mycenaean Diary.* Detroit: Wayne State University Press, 1986.

Chambers, Edmund Kerchever. *The History and Motives of Literary Forgeries.* Folcroft: Folcroft Library Editions, 1975.

Chubin, Daryl E., and Edward J. Hackett. *Peerless Science: Peer Review and U.S. Science Policy.* Albany: State University of New York Press, 1990.

Coser, Lewis A., Charles Kadushin, and Walter W. Powell. *Books: The Culture and Commerce of Publishing*. New York: Basic Books, 1982.

Council of Biology Editors. *Economics of Scientific Journals*. Bethesda, Md.: Council of Biology Editors, 1982.

Diener, Edward, and Rick Crandall. *Ethics in Social and Behavioral Research*. Chicago: University of Chicago Press, 1978.

Dutton, Denis, ed. *The Forger's Art: Forgery and the Philosophy of Art*. Berkeley, Los Angeles, London: University of California Press, 1983.

Edwards, William Allen. *Plagiarism: An Essay on Good and Bad Borrowing*. Cambridge: The Minority Press, 1933.

Erdman, David V., and Ephim G. Fogel, eds. *Evidence for Authorship: Essays on Problems of Attribution*. Ithaca: Cornell University Press, 1966.

Fried, Charles. *Right and Wrong*. Cambridge: Harvard University Press, 1978.

Garvey, William D. *Communication: The Essence of Science*. Oxford: Pergamon Press, 1979.

Gordon, Michael. *Running a Referee System*. Leicester, England: Primary Communications Research Centre, University of Leicester, 1983.

Grafton, Anthony. *Forgers and Critics: Creativity and Duplicity in Western Scholarship*. Princeton, N.J.: Princeton University Press, 1990.

Hagstrom, Warren O. *The Scientific Community*. Carbondale: Southern Illinois University Press, 1965.

Hamilton, Charles. *Great Forgers and Famous Fakes: The Manuscript Forgers of America and How They Duped the Experts*. New York: Crown Publishers, 1980.

Harrison, Wilson R. *Suspect Documents: Their Scientific Examination*. New York: Frederick A. Praeger, 1958.

Haywood, Ian. *Faking It: Art and the Politics of Forgery*. London: The Harvester Press, 1987.

Hearnshaw, L. S. *Cyril Burt, Psychologist*. London: Hodder and Stoughton, 1979.

Hering, Daniel Webster. *Foibles and Fallacies of Science; An Account of Celebrated Scientific Vagaries*. New York: D. Van Nostrand Company, 1924.

Hixson, Joseph. *The Patchwork Mouse*. Garden City: Anchor Press/Doubleday, 1976.

Horowitz, Irving Louis. *Communicating Ideas: The Crisis of Publishing in a Post-Industrial Society*. New York: Oxford University Press, 1986.

Jackson, Douglas N., and J. Philippe Rushton, eds. *Scientific Excellence: Origins and Assessment*. Newbury Park: Sage Publications, 1987.

Jahn, Melvin E., and Daniel J. Woolf, eds. *The Lying Stones of Dr. Johann Bartholomew Adam Beringer, being his Lithographiae Wirceburgensis*. Berkeley: University of California Press, 1963.

Jones, Mark, ed. *Fake? The Art of Deception*. Berkeley, Los Angeles, Oxford: University of California Press, 1990.

Kaplan, Benjamin. *An Unhurried View of Copyright*. New York: Columbia University Press, 1967.

Klaidman, Stephen, and Tom L. Beauchamp. *The Virtuous Journalist*. New York: Oxford University Press, 1987.

Klein, Alexander, ed. *Grand Deception: The World's Most Spectacular and Suc-*

cessful Hoaxes, Impostures, Ruses, and Frauds. Philadelphia: J. B. Lippincott Company, 1955.

Koestler, Arthur. *The Case of the Midwife Toad.* London: Hutchinson, 1971.

Kohn, Alexander. *False Prophets: Fraud and Error in Science and Medicine.* New York: Basil Blackwell, 1986.

Kronick, David A. *A History of Scientific and Technical Periodicals: The Origins and Development of the Scientific and Technical Press, 1665–1790,* 2d ed. Metuchen: The Scarecrow Press, 1976.

Levi, Michael. *Regulating Fraud: White-Collar Crime and the Criminal Process.* London: Tavistock Publications, 1987.

Lindey, Alexander. *Plagiarism and Originality.* New York: Harper and Brothers, 1952.

Lindsey, Duncan. *The Scientific Publication System in Social Science.* San Francisco: Jossey-Bass Publishers, 1978.

Lock, Stephen. *A Difficult Balance: Editorial Peer Review in Medicine.* Philadelphia: ISI Press, 1986.

Mallon, Thomas. *Stolen Words: Forays into the Origins and Ravages of Plagiarism.* New York: Ticknor and Fields, 1989.

Meyer, Philip. *Ethical Journalism.* New York: Longman, 1987.

Morrissette, Bruce. *The Great Rimbaud Forgery: The Affair of "La Chasse spirituelle."* St. Louis: Committee on Publications, Washington University, 1956.

Morton, A(ndrew) Q(ueen). *Literary Detection: How to Prove Authorship and Fraud in Literature and Documents.* New York: Charles Scribner's Sons, 1978.

Morton, Herbert C., and Anne J. Price, with Robert Cameron Mitchell. *The ACLS Survey of Scholars—Final Report.* Washington: American Council of Learned Societies, 1989.

Novick, Peter. *That Noble Dream: The "Objectivity Question" and the American Historical Profession.* Cambridge: Cambridge University Press, 1988.

Paull, Harry M. *Literary Ethics: A Study in the Growth of the Literary Conscience.* London: Thornton Butterworth Ltd., 1928.

Pelikan, Jaroslav. *Scholarship and Its Survival.* Lawrenceville: Princeton University Press, 1983.

Powell, Walter W. *Getting into Print: The Decision-Making Process in Scholarly Publishing.* Chicago: The University of Chicago Press, 1985.

Reagan, Charles E. *Ethics for Scientific Researchers,* 2d ed. Springfield: Charles C. Thomas, 1971.

Redfern, Walter. *Clichés and Coinages.* London: Basil Blackwell, 1989.

Rieth, Adolf. *Archaeological Fakes,* trans. Diana Imber. New York: Praeger Publishers, 1970. First published in German in 1967.

Robison, Wade L., Michael S. Pritchard, and Joseph Ellin, eds. *Profits and Professions: Essays in Business and Professional Ethics.* Clifton, N.J.: Humana Press, 1983.

Sagarin, Edward. *Deviants and Deviance: An Introduction to the Study of Disvalued People and Behavior.* New York: Praeger Publishers, 1975.

St. Onge, K. R. *The Melancholy Anatomy of Plagiarism.* Lanham, Md.: The University Press of America, 1988.

Salzman, Maurice. *Plagiarism, the "Art" of Stealing Literary Material*. Los Angeles: Parker, Stone and Baird Co., 1931.

Savan, Beth. *Science under Siege: The Myth of Objectivity in Scientific Research*. Montreal: CBC Enterprises, 1988.

Sheleff, Leon S. *The Bystander: Behavior, Law, Ethics*. Lexington, Mass.: Lexington Books, D. C. Heath and Co., 1978.

Shuchman, Hedvah L. *Self-Regulation in the Professions: Accounting, Law, Medicine*. Glastonbury, Conn.: The Futures Group, 1981.

Singleton, Alan. *Societies and Publishers: Hints on Collaboration in Journal Publishing*. Leicester, England: Primary Communications Research Centre, University of Leicester, 1980.

Small, Christopher. *The Printed Word: An Instrument of Popularity*. Aberdeen, Scotland: Aberdeen University Press, 1982.

Spencer, Frank. *Piltdown: A Scientific Forgery*. New York: Oxford University Press, 1990.

Unseld, Siegfried. *The Author and His Publisher*. Trans. Hunter Hannum and Hildegarde Hannum. Chicago: The University of Chicago Press, 1980.

Weiner, J. S. *The Piltdown Forgery*. New York: Oxford University Press, 1955; reprinted by Dover Publications, 1980.

Wilson, John T. *Academic Science, Higher Education, and the Federal Government, 1950–1983*. Chicago: The University of Chicago Press, 1983.

Wright, Christopher. *The Art of the Forger*. London: Gordon Fraser Gallery Ltd., 1984.

REPORTS

American Association for the Advancement of Science. *Scientific Freedom and Responsibility*. Washington, D.C.: American Association for the Advancement of Science, 1975.

American Association for the Advancement of Science–American Bar Association. *Project on Scientific Fraud and Misconduct, Report on Workshop Number One*. Washington, D.C.: American Association for the Advancement of Science, 1988.

———. *Project on Scientific Fraud and Misconduct, Report on Workshop Number Two*. Washington, D.C.: American Association for the Advancement of Science, 1989.

———. *Project on Scientific Fraud and Misconduct, Report on Workshop Number Three*. Washington, D.C.: American Association for the Advancement of Science, 1989.

American Association of Universities. *Framework for Institutional Policies and Procedures to Deal with Fraud in Research*. Washington, D.C.: American Association of Universities, 1988.

American Council of Learned Societies. *Scholarly Communication: The Report of the National Enquiry*. New York: American Council of Learned Societies, 1978.

Guarding the Guardians: Research on Peer Review. Proceedings of the First International Congress on Peer Review in Biomedical Publication, Chicago, 10–12 May 1989. Chicago: American Medical Association, 1989.

Institute of Medicine. *The Responsible Conduct of Research in the Health Sciences.* Washington, D.C.: National Academy Press, 1989.

"The Morality of Scientists: I—Procedures for Responding to Charges of Unethical Research Practices: University of California, San Diego, School of Medicine," and "The Morality of Scientists: II—Report of the Faculty *ad hoc* Committee to Investigate Research Fraud," *Minerva* 25 (Winter 1987): 502–512.

Office of Technology Assessment. *Federal Scientific and Technical Information in an Electronic Age: Opportunities and Challenges.* Washington, D.C.: U.S. Government Printing Office, 1989.

———. *The Regulatory Environment for Research.* Washington, D.C.: U.S. Government Printing Office, 1986.

———. *Science, Technology and the First Amendment.* Washington, D.C.: U.S. Government Printing Office, 1988.

"Policy for Dealing with Faculty Fraud in Research, University of Florida," *Minerva* 23 (Summer 1985): 305–308.

President's Commission for the Study of Ethical Problems in Medicine and Biomedical and Behavioral Research. *Whistleblowing in Biomedical Research: Policies and Procedures for Responding to Reports of Misconduct, Proceedings of a Workshop.* Washington, D.C.: U.S. Government Printing Office, 1982.

"Report of *Ad Hoc* Committee to Evaluate Research of Dr. John R. Darsee at Emory University," *Minerva* 23 (Summer 1985): 276–305.

"Report of the Committee on Academic Fraud, The University of Chicago," *Minerva* 24 (Summer–Autumn 1986): 347–358.

The Royal Society. *The Social Responsibilities of Scientists.* London: The Royal Society, 1980.

Sigma Xi. *Honor in Science.* New Haven: Sigma Xi, The Scientific Research Society, 1986.

U.S. Congress, House Committee on Science and Technology. *American Science and Policy Issues,* Chairman's Report, 99th Congress, 2d session. Washington, D.C.: U.S. Government Printing Office, 1986.

———. *The National Science Board: Science Policy and Management for the National Science Foundation, 1968–1980,* 98th Congress, 1st session. Washington, D.C.: U.S. Government Printing Office, 1983.

U.S. Congress, House Committee on Science and Technology, Subcommittee on Investigations and Oversight. *Fraud in Biomedical Research,* Hearings, 31 March 1981, 97th Congress, 1st session. Washington, D.C.: U.S. Government Printing Office, 1981.

———. *Maintaining the Integrity of Scientific Research,* Hearings, 101st Congress, 1st session. Washington, D.C.: U.S. Government Printing Office, 1990.

U.S. Congress, House Committee on Science and Technology, Task Force on Science Policy. *Research and Publications Practices* (Science Policy Study—Hearings vol. 22), 99th Congress, 2d session. Washington, D.C.: U.S. Government Printing Office, 1987.

University of Michigan, Joint Task Force on Integrity of Scholarship. *Maintaining the Integrity of Scholarship.* Ann Arbor: University of Michigan, 1984.

ARTICLES

Abelson, Philip H. "Mechanisms for Evaluating Scientific Information and the Role of Peer Review," *Journal of the American Society for Information Science* 41 (1990): 216–222.

Abrams, Floyd. "Why We Should Change the Libel Law," *New York Times Magazine* (29 September 1985).

Alexander, Brian. "Archeology and Looting Make a Volatile Mix," *Science* 250 (23 November 1990): 1074–1075.

Altman, Lawrence K. "Cheating in Science and Publishing," *CBE Views* 4 (Winter 1981): 19–25.

Altman, Philip L. "The Council of Biology Editors," *Scholarly Publishing* 20 (July 1989): 218–226.

Angell, Marcia. "Publish or Perish: A Proposal," *Annals of Internal Medicine* 104 (February 1986): 261–262.

Armstrong, J. Scott. "Research in Scientific Journals: Implications for Editors and Authors," *Journal of Forecasting* 1 (1982): 83–104.

Ayala, Francisco J. "For Young Scientists, Questions of Protocol and Propriety Can Be Bewildering," *Chronicle of Higher Education* 36 (22 November 1989): 36.

Babbage, Charles. "The Decline of Science in England," *Nature* 340 (17 August 1989): 499–502. [Reprint of 1830 original.]

Bailar, John C., III. "Science, Statistics, and Deception," *Annals of Internal Medicine* 104 (February 1986): 259–260.

Bailer, John C., III, and Kay Patterson. "Journal Peer Review: The Need for a Research Agenda," *New England Journal of Medicine* 312 (1985): 654–657.

Baltimore, David. "Baltimore's Travels," *Issues in Science and Technology* 5 (Summer 1989): 48–54.

Barber, Bernard. "Trust in Science," *Minerva* 25 (Spring–Summer 1987): 123–134.

Bechtel, H. Kenneth, Jr., and Willie Pearson, Jr. "Deviant Scientists and Scientific Deviance," *Deviant Behavior* 6 (1985): 237–252.

Benedict, Michael Les. " 'Fair Use' of Unpublished Sources," *Perspectives* 28 (April 1990): 1, 9–10, 12–13.

———. "Historians and the Continuing Controversy over Fair Use of Unpublished Manuscript Materials," *American Historical Review* 91 (October 1986): 859–881.

Bergman, Robert G. "Irreproducibility in the Scientific Literature: How Often Do Scientists Tell the Whole Truth and Nothing But the Truth?" *Perspectives on the Professions* 8 (January 1989): 2–3.

Bracey, Gerald W. "The Time Has Come to Abolish Research Journals: Too Many Are Writing Too Much about Too Little," *Chronicle of Higher Education* 30 (25 March 1987): 44–45.

Brackbill, Yvonne, and André E. Hellegers. "Ethics and Editors," *Hastings Center Report* 10 (February 1980): 20–22.

Bradley, S. Gaylen. "Ownership of Intellectual Property: Grant Applications and Technical Manuscripts," *ASM News* 52 (November 1986): 564–565.

———. "Scientific Fraud and Experimental Error," *ASM News* 55 (September 1989): 452–453.

Branscomb, Lewis M. "Integrity in Science," *American Scientist* 73 (September–October 1985): 421–423.

Braunwald, Eugene. "Commentary: On Analysing Scientific Fraud," *Nature* 325 (15 January 1987): 215–216.

Bridgstock, Martin. "A Sociological Approach to Fraud in Science," *Australian and New Zealand Journal of Sociology* 18 (November 1982): 364–383.

Broad, William J. "Crisis in Publishing: Credit or Credibility," *BioScience* 32 (September 1982): 645–647.

Broad, William J., and Nicholas Wade. "Science's Faulty Fraud Detectors," *Psychology Today* 16 (November 1982): 51–57.

Burman, Kenneth D. "'Hanging from the Masthead': Reflections on Authorship," *Annals of Internal Medicine* 97 (October 1982): 602–605.

Burnham, John C. "The Evolution of Editorial Peer Review," *Journal of the American Medical Association* 263 (9 March 1990): 1323–1329.

Carton, Barbara. "The Mind of a Whistle Blower," *Boston Globe* (1 April 1991): 42–43.

Chamberlain, A. P. "The Piltdown 'Forgery,'" *New Scientist* 40 (28 November 1968): 516.

Chubin, Daryl E. "Allocating Credit and Blame in Science," *Science, Technology, & Human Values* 13 (Winter 1988): 53–63.

———. "Misconduct in Research: An Issue of Science Policy and Practice," *Minerva* 23 (Summer 1985): 175–202.

———. "Much Ado about Peer Review," *BioScience* 36 (January 1986): 18, 20–21.

———. "Research Malpractice," *BioScience* 35 (February 1985): 80–89.

———. "A Soap Opera for Science," *BioScience* 37 (April 1987): 259–261.

———. "Values, Controversy, and the Sociology of Science," *Bulletin of Science, Technology, and Society* 1 (1981): 427–436.

Coghlan, Andy. "Researchers Roll Back the Frontiers of Fraud," *New Scientist* 113 (22 January 1987): 23.

Colt, George Howe. "Too Good To Be True," *Harvard Magazine* (July–August 1983): 22–28, 54.

Connor, Patrick E. "Scientific Research Competence," *Pacific Sociological Review* 15 (July 1972): 355–366.

Cournand, André. "The Code of the Scientist and Its Relationship to Ethics," *Science* 198 (18 November 1977): 699–705.

Cournand, André, and Michael Meyer. "The Scientist's Code," *Minerva* 14 (Spring 1976): 79–96.

Cowen, Robert. "The Not-So-Hallowed Halls of Science," *Technology Review* 86 (May–June 1983): 8, 85.

Crandall, Rick. "Peer Review: Improving Editorial Performance," *BioScience* 36 (October 1986): 607–609.

Crane, Diana. "The Gatekeepers of Science: Some Factors Affecting the Selection of Articles for Scientific Journals," *American Sociologist* 2 (November 1967): 195–201.

Crawford, Susan, and Loretta Stucki. "Peer Review and the Changing Research Record," *Journal of the American Society for Information Science* 41 (1990): 223–228.

Croll, Roger P. "The Non-Contributing Author: An Issue of Credit and Responsibility," *Perspectives in Biology and Medicine* 27 (Spring 1984): 401–407.

Davis, Bernard D. "Fraud: Why Auditing Laboratory Records Is a Bad Idea," *The Scientist* 3 (20 February 1989): 9.

Davis, Ralph D. "New Censors in the Academy," *Science, Technology, & Human Values* 13 (Winter 1988): 64–74.

Dickens, Bernard M. "Ethical Issues in Psychiatric Research," *International Journal of Law and Psychiatry* 4 (Summer–Fall 1981): 271–292.

Dixon, Bernard. "John Maddox Offers Surprising Insights into His *Nature*," *The Scientist* 2 (13 June 1988): 13–14.

Doob, Anthony N. "Understanding the Nature of Investigations into Alleged Fraud in Alcohol Research: A Reply to Walker and Roach," *British Journal of Addiction* 79 (1984): 169–174.

Edge, David. "Quantitative Measures of Communication in Science: A Critical Review," *History of Science* 17 (1979): 102–134.

Edsall, John T. "Two Aspects of Scientific Responsibility," *Science* 212 (3 April 1981): 11–14.

Engler, Robert L., James W. Covell, Paul J. Friedman, Philip S. Kitcher, and Richard M. Peters. "Misrepresentation and Responsibility in Medical Research," *New England Journal of Medicine* 317 (26 November 1987): 1383–1389.

Epstein, Richard A. "On Drafting Rules and Procedures for Academic Fraud," *Minerva* 24 (Summer–Autumn 1986): 344–347.

Fabergé, A. C. "Open Information and Secrecy in Research," *Perspectives in Biology and Medicine* 25 (Winter 1982): 263–278.

"Fraud and Secrecy: The Twin Perils of Science," *New Scientist* 98 (9 June 1983): 712–713.

Freedman, Morris. "Plagiarism among Professors or Students Should Not Be Excused or Treated Gingerly," *Chronicle of Higher Education* 34 (10 February 1988): A48.

Friedman, Paul J. "Research Ethics, Due Process, and Common Sense," *Journal of the American Medical Association* 260 (7 October 1988): 1937–1938.

Garfield, Eugene. "More on the Ethics of Scientific Publication: Abuses of Authorship Attribution and Citation Amnesia Undermine the Reward System of Science," *Current Contents*, no. 30 (26 July 1982): 5–10.

———. "What Do We Know about Fraud and Other Forms of Intellectual Dishonesty in Science? Part 1. The Spectrum of Deviant Behavior in Science," *Current Contents*, no. 14 (6 April 1987): 3–7.

———. "What Do We Know about Fraud and Other Forms of Intellectual Dishonesty in Science? Part 2. Why Does Fraud Happen and What Are Its Effects?" *Current Contents*, no. 15 (13 April 1987): 3–10.

Garrow, David J. "A Law Defining 'Fair Use' of Unpublished Sources Is Essential to the Future of American Scholarship," *Chronicle of Higher Education* 36 (18 April 1990): A48.

Gaston, Jerry. "Secretiveness and Competition for Priority of Discovery in Physics," *Minerva* 9 (October 1971): 472–492.

Glazer, Sarah. "Combating Science Fraud," *Congressional Quarterly's Editorial Research Reports* 2 (5 August 1988): 390–403.

Goellner, Jack G. "The Other Side of the Fence: Scholarly Publishing as Gate-keeper," *Book Research Quarterly* 4 (Spring 1988): 15–19.

Golley, Frank B. "Ethics in Publishing," *CBE Views* 4 (Winter 1981): 26–30.

Gould, Stephen. "The Case of the Creeping Fox Terrier Clone," *Natural History* 97 (January 1988): 16–24.

———. "The Piltdown Conspiracy," *Natural History* 89 (August 1980): 8–28.

———. "Piltdown Revisited," *Natural History* 88 (March 1979): 86–97.

Gratzer, Walter. "Trouble at t'Lab," *Nature* 302 (28 April 1983): 774–775.

Greenberg, Daniel S. "Fraud and the Scientific Method," *New Scientist* 112 (6 November 1986): 64.

———. "Part I: Chronicle of a Scientific Misconduct Case," *Science & Government Report* 16 (1 November 1986): 1–4.

———. "Part II: Chronicle of a Scientific Misconduct Case," *Science & Government Report* 16 (15 November 1986): 1–5.

———. "Part III: Chronicle of a Scientific Misconduct Case," *Science & Government Report* 16 (1 December 1986): 1–4.

———. "Scientific Fraud," *American Scientist* 72 (January–February 1984): 118.

Greene, Penelope J., Jane S. Durch, Wendy Horwitz, and Valwyn S. Hooper. "Policies for Responding to Allegations of Fraud in Research," *Minerva* 23 (Summer 1985): 203–215.

Grove, J. W. "Rationality at Risk: Science against Pseudoscience," *Minerva* 23 (Summer 1985): 216–240.

Gupta, Vishwa Jit. "The Peripatetic Fossils: Part 2," *Nature* 341 (7 September 1989): 11–12.

Hammond, Michael. "The Expulsion of the Neanderthals from Human Ancestry: Marcellin Boule and the Social Context of Scientific Research," *Social Studies of Science* 12 (February 1982): 1–36.

———. "A Framework of Plausibility for an Anthropological Forgery: The Piltdown Case," *Anthropology* 3 (1979): 47–58.

Heinz, Lisa C., and Daryl E. Chubin. "Congress Investigates Scientific Fraud," *BioScience* 38 (September 1988): 559–561.

Herbert, Wally. "Did Peary Reach the Pole?" *National Geographic* (September 1988): 387–413.

Hetherington, Norris B. "Just How Objective Is Science?" *Nature* 306 (22–29 December 1983): 727–730.

Hollis, Bruce W. "I Turned in My Mentor," *The Scientist* 1 (14 December 1987): 11–12.

Holton, Gerald J. "Do Scientists Need a Philosophy?" *Times Literary Supplement*, no. 4257 (2 November 1984): 1231–1234.

———. "Neils Bohr and the Integrity of Science," *American Scientist* 74 (May–June 1986): 237–243.

Horowitz, Irving Louis, and Mary E. Curtis. "The Impact of the New Information Technology on Scientific and Scholarly Publishing," *Journal of Information Science* 4 (1982): 87–96.

Hunt, Morton. "Did the Penalty Fit the Crime?" *New York Times Magazine* (14 May 1989): 36–37, 69–70, 73, 84, 98.

Huth, Edward J. "Guidelines on Authorship of Medical Papers," *Annals of Internal Medicine* 104 (February 1986): 269–274.

―――. "Irresponsible Authorship and Wasteful Publication," *Annals of Internal Medicine* 104 (February 1986): 257–259.

Ingelfinger, Franz J. "Peer Review in Biomedical Publication," *American Journal of Medicine* 56 (May 1974): 686–692.

International Committee of Medical Journal Editors. "Uniform Requirements for Manuscripts Submitted to Biomedical Journals," *Annals of Internal Medicine* 108 (February 1988): 258–265.

Jackson, C. Ian, and John W. Prados. "Honor in Science," *American Scientist* 71 (September–October 1983): 462–464.

Jacobstein, Jerome. "I Am Not Optimistic," *The Scientist* 1 (14 December 1987): 11–12.

―――. "Science Isn't Really Willing to Investigate Misconduct," *The Scientist* 2 (12 December 1988): 9, 12.

Jasanoff, Sheila. "Reviewing the Reviewers," *BioScience* 37, no. 4 (April 1987): 291–293.

Jones, D. G. "Guided Hand or Forgery?" *Journal of the Forensic Science Society* 26 (May–June 1986): 169–173.

Jones, F. L. "Obsession Plus Pseudo-Science Equals Fraud: Sir Cyril Burt, Intelligence, and Social Mobility," *Australian and New Zealand Journal of Sociology* 16 (March 1980): 48–55.

Kennedy, Donald. "Counting the Costs of Regulation," *The AAAS Observer*, no. 3 (6 January 1989): 5–7.

Kerr, Clark. "The Academic Ethic and University Teachers: A 'Disintegrating Profession'?" *Minerva* 27 (Summer–Autumn 1989): 139–156.

Knight, James A. "Exploring the Compromises of Ethical Principles in Science," *Perspectives in Biology and Medicine* 27 (Spring 1984): 432–442.

Knox, Richard A. "Deeper Problems for Darsee: Emory Probe," *Journal of the American Medical Association* 249 (3 June 1983): 2867–2873, 2876.

―――. "Doctor-Editor-Patient," *New England—The Boston Sunday Globe* (16 June 1977): 26–32.

―――. "The Harvard Fraud Case: Where Does the Problem Lie?" *Journal of the American Medical Association* 249 (8 April 1983): 1797–1807.

Kuznik, Frank. "Fraud Busters," *Washington Post Magazine* (14 April 1991): 22–26, 31–33.

LaFollette, Marcel C. "Beyond Plagiarism: Ethical Misconduct in Scientific and Technical Publishing," *Book Research Quarterly* 4 (Winter 1988–89): 66–73.

List, C. J. "Scientific Fraud: Social Deviance or the Failure of Virtue?" *Science, Technology, & Human Values* 10 (Fall 1985): 27–36.

Littlejohn, Marilyn J., and Christine M. Matthews. "Scientific Misconduct in Academia: Efforts to Address the Issue," Congressional Research Service, Library of Congress, Report No. 89–392 SPR (30 June 1989).

Ludbrook, John. "Plagiarism and Fraud," *Australian and New Zealand Journal of Surgery* 56 (1986): 741–742.

Lukas, Mary. "Teilhard and the Piltdown 'Hoax': A playful prank gone too far? Or a deliberate scientific forgery? Or, as it now appears, nothing at all?" *America* 144 (23 May 1981): 424–427.

Lundberg, George, and Annette Flanagan. "New Requirements for Authors: Signed Statements of Authorship Responsibility and Financial Disclosure,"

Journal of the American Medical Association 262 (13 October 1989): 2003–2005.

McBride, Gail. "The Sloan–Kettering Affair: Could It Have Happened Anywhere?" *Journal of the American Medical Association* 229 (9 September 1974): 1391–1410.

Macklin, Ruth. "Ethics in Research," *CBE Views* 4 (Winter 1981): 12–18.

McReynolds, Paul. "Reliability of Ratings of Research Papers," *American Psychologist* 26 (April 1971): 400–401.

Majerus, Philip W. "Fraud in Medical Research," *Journal of Clinical Investigation* 70 (July 1982): 213–217.

Manwell, Clyde, and C. M. Ann Baker. "Honesty in Science: A Partial Test of a Sociobiological Model of the Social Structure of Science," *Search* 12 (June 1981): 151–159.

Martin, Brian. "Fraud and Australian Academics," *Thought & Action* 52 (Fall 1989): 95–102.

Mazur, Allan. "Allegations of Dishonesty in Research and Their Treatment by American Universities," *Minerva* 27 (Summer–Autumn 1989): 177–194.

Merton, Robert K. "Scientific Fraud and the Fight to be First," *Times Literary Supplement*, no. 4257 (2 November 1984), no pages listed.

———. "Three Fragments from a Sociologist's Notebook: Establishing the Phenomenon, Specified Ignorance, and Strategic Research Materials," *Annual Reviews of Sociology* 13 (1987): 1–28.

Miers, Mary L. "Current NIH Perspectives on Misconduct in Science," *American Psychologist* 40 (July 1985): 831–835.

Minsky, Leonard. "Fraud Is a Symptom of a Deeper Flaw," *The Scientist* 2 (12 December 1988): 9, 12.

Mishkin, Barbara. "Responding to Scientific Misconduct: Due Process and Prevention," *Journal of the American Medical Association* 260 (7 October 1988): 1932–1936.

Muller, Mike. "Why Scientists Don't Cheat," *New Scientist* 74 (2 June 1977): 522–523.

"NIH's Raub on Misconduct," *The Scientist* 1 (15 December 1986): 18–19.

Nissenbaum, Stephen. "The Plagiarists in Academe Must Face Formal Sanctions," *Chronicle of Higher Education* 36 (28 March 1990): A52.

Okie, Susan. "When Researchers Disagree," *Washington Post* 111 (11 April 1988): A1, A12.

O'Neil, Robert M. "Scientists Accused of Fraud Deserve Procedural Safeguards," *Chronicle of Higher Education* 37 (28 November 1990): A48.

O'Toole, Margot. "Scientists Must Be Able to Disclose Colleagues' Mistakes Without Risking Their Own Jobs or Financial Support," *Chronicle of Higher Education* 34 (25 January 1989): A44.

Owens, R. Glynn, and E. M. Hardley. "Plagiarism in Psychology—What Can and Should Be Done?" *Bulletin of the British Psychological Society* 38 (October 1985): 331–333.

Parsons, Paul F. "The Editorial Committee: Controller of the Imprint," *Scholarly Publishing* 20 (July 1989): 238–244.

Petersdorf, Robert G. "The Pathogenesis of Fraud in Medical Science," *Annals of Internal Medicine* 104 (February 1986): 252–254.

Pigman, Ward, and Emmett B. Carmichael. "An Ethical Code for Scientists," *Science* 111 (16 June 1950): 643–647.

Pinch, Trevor J. "Normal Explanations of the Paranormal: The Demarcation Problem and Fraud in Parapsychology," *Social Studies of Science* 9 (August 1979): 329–348.

Pollock, A. V., and Mary Evans. "Bias and Fraud in Medical Research: A Review," *Journal of the Royal Society of Medicine* 78 (November 1985): 937–940.

Powledge, Tabitha M. "NEJM's Arnold Relman," *The Scientist* 2 (21 March 1988): 12–13.

———. "Stewart and Feder (Finally) in Print," *The Scientist* 1 (9 February 1987): 12.

Racker, Efraim. "A View of Misconduct in Science," *Nature* 339 (11 May 1989): 91–93.

Rennie, Drummond. "How Much Fraud? Let's Do an Experimental Audit," *AAAS Observer*, no. 3 (6 January 1989): 4.

Reti, Ladislao. "Francesco di Giorgio Martini's Treatise on Engineering and Its Plagiarists," *Technology and Culture* 4 (Summer 1963): 287–298.

Roman, Mark B. "When Good Scientists Turn Bad," *Discover* (April 1988): 50–58.

Rosen, Dennis. "The Jilting of Athene," *New Scientist* 39 (5 September 1968): 497–500.

Rosenzweig, Robert M. "Preventing Fraud Is a Task for Scientists, Not Congress," *The Scientist* 2 (12 December 1988): 9, 12.

Rothberg, Morey D. "'To Set a Standard of Workmanship and Compel Men to Conform to It': John Franklin Jameson as Editor of the *American Historical Review*," *American Historical Review* 89 (October 1984): 957–975.

Sabine, John R. "The Error Rate in Biological Publications: A Preliminary Survey," *Science, Technology, & Human Values* 10 (Winter 1985): 62–69.

Sachs, Harley L. "The Publication Requirement Should Not Be Based Solely on Refereed Journals," *Chronicle of Higher Education* 35 (19 October 1988): B2.

St. James-Roberts, Ian. "Are Researchers Trustworthy?" *New Scientist* (2 September 1976): 481–483.

———. "Cheating in Science," *New Scientist* (25 November 1976): 466–469.

Sanders, Howard J. "Peer Review: How Well Is It Working?" *Chemical & Engineering News* 60 (15 March 1982): 32–43.

Schaefer, William D. "Much 'Scholarship' in the Humanities Is Done Badly and Probably Shouldn't Be Done at All," *Chronicle of Higher Education* 36 (7 March 1990): B1–B2.

Schmaus, Warren. "Fraud and the Norms of Science," *Science, Technology, & Human Values* 8 (Fall 1983): 12–22.

Segerstrale, Ullica. "'Right Is Might': Physicists on Fraud, Fudging, and 'Good Science,'" *Perspectives on the Professions* 8 (January 1989): 5–6.

Shapiro, Martin F., and Robert P. Charrow. "The Role of Data Audits in Detecting Scientific Misconduct," *Journal of the American Medical Association* 261 (5 May 1989): 2505–2511.

Sharp, D. "La Fraude dans les Sciences: Que Peuvent Faire les Responsables de Revues," *Nouvelle Revue de Medicine de Toulouse* 2 (1984): 335–338.

Shaw, Peter. "Plagiary," *American Scholar* 51 (Summer 1982): 325–337.

Shils, Edward. "The Morality of Scientists," *Minerva* 23 (Summer 1985): 272–275.

Sisler, William P. "Loyalties and Royalties," *Book Research Quarterly* 4 (Spring 1988): 12–14.

Smigel, Erwin O., and H. Laurence Ross. "Factors in the Editorial Decision," *American Sociologist* 5 (February 1970): 19–21.

Sprague, Robert L. "I Trusted the Research System," *The Scientist* 1 (14 December 1987): 11–12.

"Stanford's Donald Kennedy: The View from the University," *The Scientist* 3 (20 February 1989): 13.

Steneck, Nicholas H. "The University and Research Values," *Science, Technology, & Human Values* 9 (Fall 1984): 6–15.

Stewart, Walter W., and Ned Feder. "The Integrity of the Scientific Literature," *Nature* 325 (15 January 1987): 207–214.

———. "We Must Deal Realistically with Fraud and Error," *The Scientist* 1 (14 December 1987): 13.

———. "Why Research Fraud Thrives," *Boston Sunday Globe* 230 (30 November 1986): A21, A24.

Stossel, Thomas P. "SPEED: An Essay on Biomedical Communication," *New England Journal of Medicine* 313 (11 July 1985): 123–126.

Suttmeier, Richard P. "Corruption in Science: The Chinese Case," *Science, Technology, & Human Values* 10 (Winter 1985): 49–61.

Swazey, Judith P., Karen Seashore Louis, and Melissa S. Anderson. "University Policies and Ethical Issues in Research and Graduate Education: Highlights of the CGS Deans' Survey," *CGS Communicator* 22 (March 1989): 1–3, 7–8.

Sweeting, Linda M. "Chemistry's Toxic Image," *The Scientist* 1 (15 December 1986): 11.

Szilagyi, D. Emerick. "The Elusive Target: Truth in Scientific Reporting—Comments on Error, Self-Delusion, Deceit, and Fraud," *Journal of Vascular Surgery* 1 (March 1984): 243–253.

Tangney, June Price. "Fraud Will Out—or Will It?" *New Scientist* 115 (6 August 1987): 62–63.

Thompson, Larry. "Science under Fire," *Health* [supplement to the *Washington Post*] 5 (9 May 1989): 12–16.

"Trials and Tribulations of Being Director of the NIH," *The Scientist* 2 (11 July 1988): 13–14.

Veliotes, Nicholas. "Copyright in the 1990s: A New Round of Challenges for American Publishers," *Book Research Quarterly* 4 (Spring 1988): 3–11.

Wade, Nicholas. "Madness in Their Method," *New Republic* 188 (27 June 1983): 13–17.

———. "The Rise and Fall of a Scientific Superstar," *New Scientist* 91 (24 September 1981): 781–782.

Weinstein, Deena. "Fraud in Science," *Social Science Quarterly* 59 (March 1979): 638–652.

Weiss, Ted. "Research That the U.S. Government Is Paying For Should Not Be Tainted by any Possibility of Bias," *Chronicle of Higher Education* 36 (4 October 1989): A56.

———. "Too Many Scientists Who 'Blow the Whistle' End Up Losing Their Jobs and Careers," *Chronicle of Higher Education* 37 (26 June 1991): A36.

West, Susan. "Crackdown on Lab Fraud," *Science 83* (May–June 1983): 14.

Woolf, Patricia. "Deception in Scientific Research," *Jurimetrics Journal* 29 (Fall 1988): 67–95.

Woolf, Patricia. "Fraud in Science: How Much, How Serious?" *Hastings Center Report* 11 (October 1981): 9–14.

———. "How Ethical Is Science?" *CBE Views* 4 (Winter 1981): 5–11.

———. "Pressure to Publish and Fraud in Science," *Annals of Internal Medicine* 104 (February 1986): 254–256.

———. "'Pressure to Publish' Is a Lame Excuse for Scientific Fraud," *Chronicle of Higher Education* 34 (23 September 1987): A52.

———. "Science Needs Vigilance Not Vigilantes," *Journal of the American Medical Association* 260 (7 October 1988): 1939–1940.

Yankauer, Alfred. "Peering at Peer Review," *CBE Views* 8 (Summer 1985): 7–10.

Zuckerman, Harriet. "Intellectual Property and Diverse Rights of Ownership in Science," *Science, Technology, & Human Values* 13 (Winter–Spring 1988): 7–17.

———. "Norms and Deviant Behavior in Science," *Science, Technology, & Human Values* 9 (Winter 1984): 7–13.

———. "Patterns of Name Ordering among Authors of Scientific Papers: A Study of Social Symbolism and Its Ambiguity," *American Journal of Sociology* 74 (1968): 276–291.

Zuckerman, Harriet, and Robert K. Merton. "Patterns of Evaluation in Science: Institutionalisation, Structure, and Functions of the Referee System," *Minerva* 9 (January 1971): 66–100.

Zurer, Pamela S. "Misconduct in Research: It May Be More Widespread than Chemists Like to Think," *Chemical & Engineering News* 65 (13 April 1987): 10–17.

Zylke, Jody W. "Investigation Results in Disciplinary Action Against Researchers, Retraction of Articles," *Journal of the American Medical Association* 262 (13 October 1989): 1910–1911, 1915.

EDITORIALS

"An Outbreak of Piracy in the Literature," *Nature* 285 (12 June 1980): 429–430.

Bunnett, Joseph F. "Coauthorship of Research Papers," *Accounts of Chemical Research* 18 (March 1985): 63.

———. "Errors in Publications," *Accounts of Chemical Research* 18 (October 1985): 291.

"Do Scientific Journals Need a Code of Practice?" *Nature* 258 (6 November 1975): 1.

"Fraud, Libel, and the Literature," *Nature* 325 (15 January 1987): 181–182.

Garfield, Eugene. "Citation Indexes Can Help Halt the Spread of Fraudulent Research," *The Scientist* 3 (7 August 1989): 12.

———. "Contrary to *Nature*?" *The Scientist* 2 (5 September 1988): 12.

———. "Is There Room in Science for Self-Promotion?" *The Scientist* 1 (14 December 1987): 9.

———. "Journal Editors Owe Readers Timely Action on Retractions," *The Scientist* 3 (6 February 1989): 10.

———. "Opting Out of the Numbers Game," *The Scientist* 1 (23 February 1987): 9.

———. "Why Scientific Publishing Should Be Audited," *The Scientist* 3 (24 July 1989): 12.

———. "Why *The Scientist* Welcomes Corrections," *The Scientist* 2 (25 July 1988): 12.

Glaze, William H. "Peer Review: A Foundation of Science," *Environmental Science & Technology* 22 (March 1988): 235.

———. "Peer Review Expertise," *Environmental Science & Technology* 22 (December 1988): 1371.

Grouse, Lawrence D. "Dealing with Alleged Fraud in Medical Research," *Journal of the American Medical Association* 248 (1 October 1982): 1637–1638.

Heylin, Michael. "The Balm of Humility," *Chemical & Engineering News* 69 (8 April 1991): 3.

———. "Credibility," *Chemical & Engineering News* 67 (22 May 1989): 3.

———. "Credibility and Fraud," *Chemical & Engineering News* 65 (13 April 1987): 3.

———. "Science and the Great Unwashed," *Chemical & Engineering News* 66 (11 July 1988): 3.

Huth, Edward J. "Abuses and Uses of Authorship," *Annals of Internal Medicine* 104 (February 1986): 266–267.

———. "Authorship from the Reader's Side," *Annals of Internal Medicine* 97 (October 1982): 613–614.

———. "Responsibilities of Co-Authorship," *Annals of Internal Medicine* 99 (August 1983): 266–267.

Koshland, Daniel E., Jr. "The Addictive Personality," *Science* 250 (30 November 1990): 1193.

———. "Conflict of Interest," *Science* 249 (13 July 1990): 109.

———. "Fraud in Science," *Science* 235 (9 January 1987): 141.

———. "The Process of Publication," *Science* 245 (11 August 1989): 573.

———. "Science, Journalism, and Whistle-Blowing," *Science* 240 (29 April 1988): 585.

———. "The Underrepresentation Syndrome," *Science* 245 (28 July 1989): 341.

LaFollette, Marcel C. "Journal Peer Review and Public Policy," *Science, Technology, & Human Values* 10 (Winter 1985): 3–5.

———. "On Trust," *Science, Technology & Human Values* 12 (Summer–Fall 1987): 1–2.

Lewin, Benjamin. "Fraud in Science: The Burden of Proof," *Cell* 48 (16 January 1987): 1–2.

Maddox, John. "Can Journals Influence Science?" *Nature* 339 (29 June 1989): 657.

———. "How to Say Sorry Graciously," *Nature* 338 (2 March 1989): 13.

———. "Plagiarism, Piracy and Principles," *Nature* 286 (28 August 1980): 831–832.

———. "Privacy and the Peer-Review System," *Nature* 312 (6 December 1984): 497.

———. "A Too-Polite Silence about Shoddy Science," *New York Times* 88 (26 September 1988): A23.

———. "Whose Misconduct?" *Nature* 326 (30 April 1987): 814.

———. "Why the Pressure to Publish?" *Nature* 333 (9 June 1988): 493.

Meadow, Charles T. "Introduction and Overview," *Journal of the American Society for Information Science* 41 (1990): 214–215.

"Must Plagiarism Thrive?" *British Medical Journal* 281 (5 July 1980): 41–42.

Newell, Frank W. "Fabricated Data and Coauthors," *American Journal of Ophthalmology* 103 (January 1987): 101–102.

Nicholson, Richard S. "On Being a Scientist," *Science* 246 (20 October 1989): 305.

"Plagiarism and the Basic Values," *Washington Post* (1 December 1988).

"Plagiarism Strikes Again," *Nature* 286 (31 July 1980): 433.

Relman, Arnold S. "Lessons from the Darsee Affair," *New England Journal of Medicine* 308 (9 June 1983): 1415–1417.

Relman, Arnold S., and Marcia Angell. "How Good is Peer Review," *New England Journal of Medicine* 321 (21 September 1989): 827–829.

Rennie, Drummond. "Editors and Auditors," *Journal of the American Medical Association* 261 (5 May 1989): 2543–2545.

———. "Guarding the Guardians: A Conference on Editorial Peer Review," *Journal of the American Medical Association* 256 (7 November 1986): 2391–2392.

"Scientific Fraud," *Washington Post* (28 March 1991).

"Scientific Fraud: In Bristol Now," *Nature* 294 (10 December 1981): 509.

"Should Journal Editors Play Science Cops?" *The Scientist* 2 (31 October 1988): 3, 26.

Wade, Nicholas. "Fraud and Garbage in Science," *New York Times* (29 January 1987).

"The Whistle Blower on Fraud at MIT," *Boston Globe* (25 March 1991).

"Whose Misconduct?" *Nature* 326 (30 April 1987): 814.

Index

Designer: U.C. Press Staff
Compositor: Braun-Brumfield, Inc.
Text: 10/13 Aldus
Display: Aldus
Printer: Braun-Brumfield, Inc.
Binder: Braun-Brumfield, Inc.

5973